The Flow Sys

The Evolution of Agile and Lean Thinking in an Age of Complexity

By

John R. Turner, PhD • Nigel Thurlow • Brian Rivera

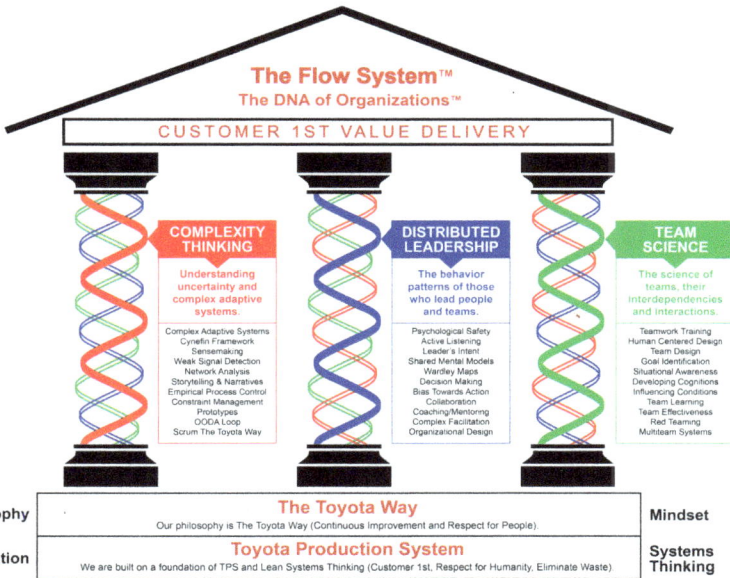

The Flow System™
The DNA of Organizations™

CUSTOMER 1ST VALUE DELIVERY

COMPLEXITY THINKING	DISTRIBUTED LEADERSHIP	TEAM SCIENCE
Understanding uncertainty and complex adaptive systems.	The behavior patterns of those who lead people and teams.	The science of teams, their interdependencies and interactions.
Complex Adaptive Systems	Psychological Safety	Teamwork Training
Cynefin Framework	Active Listening	Human Centered Design
Sensemaking	Leader's Intent	Team Design
Weak Signal Detection	Shared Mental Models	Goal Identification
Network Analysis	Wardley Maps	Situational Awareness
Storytelling & Narratives	Decision Making	Developing Cognitions
Empirical Process Control	Bias Towards Action	Influencing Conditions
Constraint Management	Collaboration	Team Learning
Prototypes	Coaching/Mentoring	Team Effectiveness
OODA Loop	Complex Facilitation	Red Teaming
Scrum The Toyota Way	Organizational Design	Multiteam Systems

Our Philosophy — **The Toyota Way** — Our philosophy is The Toyota Way (Continuous Improvement and Respect for People). — **Mindset**

Our Foundation — **Toyota Production System** — We are built on a foundation of TPS and Lean Systems Thinking (Customer 1st, Respect for Humanity, Eliminate Waste). — **Systems Thinking**

3 Helix Publishing[TM]

ISBN: 979-8-9880239-0-6 (paperback)
DOI: https://doi.org/10.12794/sps.flow-058-8

Dedications

To our readers, may you find value and benefit from the materials provided in this book.

To Jana, for your endless support and belief along the way. To Clark, I hope to be an inspiration to you as you mature. Mom, thanks.

—John R. Turner

For those who cared and believed in me. I know it was hard. To Jessica and Elliott, for being all I am not. To Jodie, for your true love. For my Dad, I miss you. Mum, I'm glad you got to see this. And okay, Sis, thanks to you, too!

—Nigel Thurlow

To my wife, Allison, my daughters, Carmen and Camille, and my parents. Allison, you make me a better person and father. To those on eternal patrol.

—Brian Rivera

Foreword

By Professor Dave Snowden, creator of the Cynefin Framework,
chief scientific officer of Cognitive Edge, and director of the Cynefin Centre

W e live in complex times, verging on the chaotic. The coronavirus pandemic has, at least in the short to medium term, radically changed the way we view the world and the body politic. A former British prime minister once coined the phrase "There is no magic money tree" to justify an economic policy of austerity, but faced with this crisis of unprecedented scale, entire *magic money forests* suddenly have cropped up in most national states. It is a truism to say that nothing will be the same again, but the nature of the difference is far from clear. Will we genuinely rethink the nature of our interactions with our fellow human beings and, more important, the planet, or will we simply lurch back to some dystopian populism and await the next crisis? The only thing that is certain about highly complex situations is that all interventions will produce unintended consequences both good and bad. Our ability to disrupt the bad and exploit the good is a measure of our resilience at all levels.

The Flow System's Triple Helix of Flow provides many of the tools and ways of thinking we will need to do that. It is *agile* without being doctrinaire about *Agile,* and its Toyota origins show in that it doesn't reject what we have done in the past, but rather finds the boundaries of those practices and then identifies new approaches for when those boundaries have to be crossed. Rethinking the nature of teams, distributing leadership, and understanding the science of uncertainty in human systems, anthro-complexity shows a willingness to go beyond a single propriety method to one that collects different methods, tools, and philosophies that are diverse enough to create resilience, but coherent enough to give an adaptive sense of direction. In anthro-complexity, we start journeys with a sense of direction, open to novel discoveries and adaption as the nature of the system with which we are dealing unfolds. This is a paradigm shift from the engineering approach that has dominated the past few decades in which futile attempts have been made to eliminate rather than embrace uncertainty. Following are some of the fairly easily understood heuristics of managing in uncertain times:

1. Centralize coordination and constraint management but distribute decision making. You simply cannot control the diversity of decisions that have to be made, and made in shortened time scales, so building trust into teams and into process will allow you to do this; coordination is the role of leadership and the ability to sense weak signals and quickly allocate resource to where it is most needed.
2. Communicate by engagement, and use your employees and networks as a distributed human sensor network to increase the diversity not only of your situational assessment but also of future scenarios. The familiar platitudes of employee engagement do not apply, if they ever did. Your staff are *entangled* with the world, your organization, and its future. Just as bramble bushes in a thicket[1] attempting to untangle what has emerged over time will simply break things, it is better to take advantage of its connections—you don't have to understand all of its many threads to take advantage of the fruit it bears.
3. Map and manage the constraints in play and understand that *enabling constraints* can reduce the cost of energy flows within the organization. Some constraints are dark, and you can see the impact but can't see the source; some constraints can be changed; and others, at least for the moment, are a given. Recognize the reality of what you can change, and more critically, identify the areas you can monitor the impact of change and rapidly redeploy resources to amplify the good and disrupt the bad.

Of course, there are many things you should not do, but to enumerate those would take a book in its own right. Two points are critical:

1. Avoid using the occasion to peddle a single solution or idea. Most of what we have done in the past has value, it was just not universal in its nature, and your bright new shiny idea will have its limits. Use what worked before up to the boundaries of its applicability and then experiment (without ideological fervor) on the other side of that boundary.
2. Don't ignore context. I've lost count over the years of the number of times case studies of successful organizations were used, and abused, to justify some new management theory. Understand the basics of biology: the first into a new sector is likely to be the apex predator and the rest of the food chain will organize around it. The success of any one entity may mask inefficiencies that will be exposed as the context shifts. You have your own path to follow. Learn from others but do not use the retrospective coherence of an airport book to constrain you to a recipe, or worse yet, a set of platitudes.

---

1 This phrase originally was used by Alicia Juarrero to describe a complex system.

The Flow System will be the start of a new journey for organizations that adopt it. It is a multimethod approach that allows you to learn but not imitate. It is based on both sound theory and rigorous practice, and in that context, it seems appropriate to end with my favorite quote from Lincoln's address to Congress on December 1, 1862:

> The dogmas of the quiet past, are inadequate to the stormy present. The occasion is piled high with difficulty, and we must rise—with the occasion. As our case is new, so we must think anew, and act anew.

If you do nothing else after reading this, note the combination: we have to both act and critically think in different ways. *The Flow System* provides an opportunity to do exactly that.

—Dave Snowden, April 2020

Foreword

By Stephen Denning, author of The Age of Agile

The coronavirus crisis has vast biological, economic, political, social, and moral implications that will be felt for years, if not decades. It will generate new ways of working, new ways of playing, and new ways of learning. It will require new ways of leading and managing. It will nurture new ways of living. Getting back to "the way we were" will no longer be possible. Everything will be different. This is the mother of all disruptions. It is ushering in a new age.

The crisis is thus proving to be a Great Accelerator. The crisis is speeding up some negative trends. Aspiring authoritarians are using the emergency to test the limits of civil rights and to consolidate power at the expense of democracy. Public bailouts of those affected risk being diverted for political purposes. Civil rights including privacy are also at risk.

Yet the crisis is also accelerating positive changes that were already underway, including the shift to digital and virtual work and learning, enhanced international collaboration, emergent leadership at the periphery, public service contributions in a spirit of solidarity, and above all, an acceleration of organizational adaptation and agility.

To date, performance varies widely. In the public sector, different governments and agencies have exhibited strikingly different degrees of dexterity, ranging from bureaucratic ineptitude to impressive quickness. The differences in adaptability have led to markedly different outcomes in terms of the impact of the disease in both biological and economic terms. Yet early gains by strong performers will be easily lost if vigilance slips.

In the private sector, business agility, which has been a major organizational trend in the past few years, has become an overriding priority. Two year ago, the best agile firms were seen to be "eating the world."[2]

Today, that dominance has become even more pronounced, as firms that had mastered more agile ways of working digitally and virtually, are able to prosper even during deep crises. For example, the share of S&P 500 market capitalization of the five largest stocks—Microsoft, Apple, Amazon, Alphabet, and Facebook—has soared to all-time highs, easily surpassing that of the dot-com bubble almost 20 years ago, and outpacing those of their weaker rivals. The market value of just two of these companies, Amazon and Microsoft, is now higher than that of the entire stock market of the United Kingdom.[3]

The Flow System is a practical guide to surviving and thriving in these times of crisis. It explains what's involved in acting with the agility of these market leaders. It shows how agile firms are succeeding in the first instance, not so much by what they do, but rather how they think, in ways that are quite different from the ponderous, top-down bureaucracies that dominated the 20th century. It reveals the principles that are driving the success of these companies, not just the practices that they implement.

It also shows how, in this new way of working, leadership is distributed throughout the organization, allowing for those closest to the problem to be capable of making decisions and taking action to meet the demands of any threats they may encounter and take advantage of any opportunities that might add value to customers. Executives provide strategic direction while pushing operational decisions to levels as close as possible to the problem or customer.

The Flow System shows how to generate and nurture self-organizing teams that mobilize the full talents of those doing the work to cope with dizzying change and complexity, while also drawing on the contributions of those for whom the work is being done—the customers.

Overall, *The Flow System* provides an essential handbook for leading and managing in these dire times.

—Stephen Denning, April 30, 2020

---

2 S. Denning, "Why Agile Is Eating the World," Forbes.com, January 2, 2018, https://www.forbes.com/sites/stevedenning/2018/01/02/why-agile-is-eating-the-world%E2%80%8B%E2%80%8B/#3899aade4a5b

3 L. Hatheway and A. Friedman, "What the Market Is Really Saying," *Project Syndicate*, April 24, 2020, https://www.project-syndicate.org/onpoint/what-the-stock-market-is-really-saying-by-larry-hatheway-and-alexander-friedman-2020-04

Foreword to *The Flow System*

By Steven Spear, author of The High Velocity Edge.

C ontrol theorists will reduce their concerns to one: cycle time. Get your detection and adaptation times within the frequencies of the environment in which you're operating, and you can thrive. Conversely, if your detection and adaptation times are too slow, you'll be overtaken by the situations you're in, sometimes with terrible consequences.

That getting adaptations faster and ensuring they're better targeted is self-evident, this being written during the throes of the novel-corona virus pandemic. This disease, compared to its compatriots SARS and MERS has spread farther, faster, and with more devastating impact. To scale that statement, SARS was measured by infections in the thousands and deaths in the hundreds. Covid-19's impact is measured by the millions infected and the hundreds of thousands killed.

Why such differences in morbidity and mortality? Cycle time. On the one hand, Covid-19 is fast. Compared to even the early 2000s, disease can move further faster because of speed and interconnectedness of transportation, so we have less time to detect and react. And once this virus lands someplace and gets a toehold in one host, it might be infectious quicker than its peers (the definitive science is still being developed), so it can spread fast and far.

On the flip side, society is slow. For a host of bureaucratic and political rea-sons, local authorities in Wuhan were plodding in sounding the alarm (though local authorities weren't necessarily faster with SARS about 20 years ago), and authorities elsewhere were slow to pay attention to the alarm. Measures to break transmission channels like limitations on flying, social distancing, and business closings that we now realize as necessary, seemed outlandish overreactions even two to three months ago. Making matters worse, it's just slow to develop tests (so we can know faster who should be isolated for a period and who needn't be) and slow too to develop vaccines and treatments.

That creates a predicament. We could ask the world to slow down, but it won't. Viruses will appear when they want to appear, and we've learned the enormous economic, social, and emotional costs of decelerating society. The alternative then is to figure out how to speed up, get our adaptation times within the cycle times of what is going on around us. Doing so means operating in situations that are increasingly non-linear and more complex (A doesn't connect to B which connects to C, and so forth. A connects to B and C and tau and rho, to several numbers, and to a variety of geometric shapes), situations which are also faster moving.

Orienting and acting in such environments means first keeping tightly aligned on our mission and purpose. Customer first. It might have been fine with simpler and slower moving systems to identify our organization's purpose once (and revalidate infrequently). That done, we were liberated to perform our roles, confident that so long as we continued, more or less, to do what we'd been doing, all would be fine.

No longer so. Who are our customers, what do they want, how can those needs be met…those questions keep recurring as today's answers are perishable. That creates a double challenge. If what we need to do keeps changing, then we have to keep reconfiguring the systems we use to anticipate, track, and meet those needs. But, those very systems will become increasingly complex and dynamic.

The term 'command and control' is often used to described management systems of the 50s, 60s, and 70s. I'm skeptical if there ever was perfect fidelity to the ideas, with some few people actually doing the thinking at the top and some many people actually following instructions at the bottom. We know that even when corporate leaders thought they were commanding and controlling, there were countless 'kluge' solutions being used on the shop floor, hence the emergence of the phrase "the hidden factory." In more authoritarian societies, the 'hidden factories' within organizations were overlaid with black markets for the economy as a whole.

Nevertheless, the notion of command and control (with its unofficial but absolutely necessary offset of local adaptation) might have been good enough at some time. Not today. Overwhelming is the evidence that companies and governments that stuck to such legacy approaches failed as everything around them sped up.

So, what's the alternative? Distributed leadership. And by this, the meaning is not everyone gets to do whatever the heck they want. That'd be chaotic. Rather, it means that everyone has a span of responsibility over which they have authority and resources to regularly detect aberration; conceive and test alternative solutions; and take corrective actions. With that combination of responsibility, authority, and rigor the component pieces of a larger system can retain their integrity and agility.

That's necessary, but insufficient. Also necessary is to ensure that the individual teams are continuously aligned and realigned into a team of teams. This means understanding the larger purpose of the system of which they are part (e.g., "customer first") and how their portion fits into the larger whole.

Organizations that pull off this triple helix trick of thinking about the complexity of their systems and the environment in which they're operating, distributing leadership to engage the collective intelligence and creativity of the organization, and building teams of teams so the whole is greater than the sum of the parts have a good chance of keeping up and staying ahead. Those who don't…well, let's try not to be one of them.

With best wishes for success,
Dr. Steve Spear DBA MS
HVE LLC, Principal
MIT Sloan School, Sr. Lecturer
The High Velocity Edge, author
www.SeeToSolve.com

Foreword to *The Flow System*

By John Shook, Chairman of the Lean Global Network *and Senior Advisor to the* Lean Enterprise Institute.

The Flow System takes on an ambitious and timely task. Organizations today face changes at a scale and speed unseen since the birth of the modern industrial enterprise. Forces of change bombard businesses from unpredictable directions. Technological changes beget challenges alongside enticing opportunities. Customer demands swing wildly and rapidly. Yet in one seemingly inevitable trajectory toward more excellent individual choice, a giant Freudian id is taking over the psyche of global consumers screaming, "I want what I want when I want it!"

Theorists proclaim the rise of organizations so endlessly flexible as to be rendered amorphous. When does total flexibility become formless? And can a formless organization provide a customer with anything of meaningful value? Anything that is physically complex and might actually work. An iPhone is a complex product and usually works phenomenally well. Usually. A Camry is even more complex and is as reliable as…a Corolla. Most apps on my phone work most of the time. The brakes on my Toyota won't fail for years, and they'll telegraph their deterioration well before they do. Thankfully.

While some apps on my phone don't always work, when they do, they work exceptionally well. They are also continually updated with new features! Most mornings, I wake up to discover upgrades that arrived magically in the night on one app or another, or even the whole operating system itself. But, unfortunately, and all too often, each step forward is accompanied by one or two in reverse. No wonder my friends at the exemplary agile software company *Menlo Innovations* proclaim as their mission to "Relieve the world of suffering caused by technology."

So, can I please have both? Can I have both reliability – in the form of things and services that work as intended – and flexibility or adaptive agility? Marrying two worlds is often hailed as necessary and immanent, but in reality, the path to change is never smooth. Everyone wants a roadmap, but roadmaps don't exist where no one has traveled. Traversing uncharted territory requires a compass and skill in finding your direction. Digital transformation, agile software development, scrum development methods, lean startup, holocracy/matrixed organization – these are all exciting concepts and practices. Sometimes it is unclear whether they are attempts to provide roadmaps or conceived of as destinations themselves.

The authors of The Flow System thankfully make no claims at offering a roadmap. Instead, they suggest a set of concepts and practices to serve as invaluable aids in navigating the uncharted waters ahead. Many of the concepts may sound "agile-ish," but the authors caution explicitly: "This is not a book about agile." We have plenty of those already.

They could have added, "This is not a book about lean." While The Flow System retains the Toyota Production System as its foundation, the authors' focus is less on seeking avenues of connecting TPS with TFS – though ample opportunities exist to draw parallels. Instead, the focus is on identifying the mindset, the sets of skills, and tools that may serve as enablers for the hapless traveler aboard organizations in the 2020s.

The authors do, however, recognize TPS and the Toyota Way as their foundation. "Lean Production" or "Lean Thinking," the reader should recall, is the label applied by researchers at MIT to Toyota's way of working in an attempt to extrapolate the principles and practices that could benefit any organization. If choosing the term "lean" was a relatively casual selection, they might well have chosen the term "flow" instead.

The visual depiction of The Flow System in the form of a house with its triple helix represented as three pillars gets its inspiration from the famous "TPS House" with its two pillars of Jidoka and JIT. The JIT pillar is, in fact, sometimes referred to as "the flow pillar." Jidoka, meanwhile, is Toyota's statement from the early 20th Century on the complex issues revolving around the relationship of humans and technology, issues also tackled by the authors of The Flow System.

In the early 21st Century, Toyota reconceived its system as the Toyota Way 2001. Defined broadly, the two pillars of the Toyota Way find easy applicability anywhere, anytime: surely, even in today's and future organizations *Continuous Improvement* and *Respect for Humanity* will still apply!

Toyota has maintained that just as its principles will surely remain constant, the infinite dimensions of *how* to actualize Continuous Improvement & Respect for People will just as surely forever be in a state of flux.

Hence, The Flow System is proposed not as a single solution but as a way forward for organizations in this new era of increasing uncertainty:

> "We are now living in complex and ambiguous times. While we understand that The Flow System is not the only source for necessary tools and techniques, we feel that The Flow System could be a starting point for many."

"We are hopeful that The Flow System, the integration of the Triple Helix of Flow, and the methods, techniques, and tools provided in The Flow System will aid everyone during this new journey.

"The act of experimentation is essential for organizations/institutions to survive in complex environments. The Flow System™ acts as a guide for organizations/institutions to focus on during these experimentations."

Indeed, we require experimentations to learn our way forward. Models and frameworks and lenses can help us make sense of the world and how we choose to interact with it, but as the statistician George Box cautioned, "All models are wrong, some are useful."

The Toyota Production System has proved to be an extraordinarily useful model for the past 70 years. Its value is confirmed only by its usefulness. Time and experience will tell, but I hope that The Flow System will prove equally useful in the years ahead. We can learn only by accepting the authors' invitation to participate in the experiment.

Foreword to *The Flow System*

By Risuo Shingo, President of the Institute of Management Improvement, *former President of* Toyota China *and* GAC Hino Motors.

I was recently wondering about the Toyota Production System when I was asked "How do you explain to the person not familiar with TPS, what TPS is?".

My answer was, it is an accumulation of small ideas. TPS means flexibility and adaptability. Team activity rather than individual activity.

TPS refers to:

- KANBAN
- 4S
- Standardization.
- Elimination of waste (MUDA).
- Continuous Improvement (KAIZEN).
- Problem Solving (Ask Why 5 Times).
- Pull System.
- Just in Time (JIT): Necessary thing/Necessary volume/Necessary timing
- Visual Management.
- Prevention of careless mistake (Poka-yoke).
- One-Piece Flow.
- Built-in-Quality.
- Jidoka (Automation with human wisdom) Detect the machine problem/stop the machine/notify the worker.
- Levelling Production (Heijunka).

Concerning Leadership:

- Leadership is very important factor to be successful.
- Show the people the target to achieve.
- Show your back (Lead by example).

- Never give up.
- Big Ear (Listen carefully) Big Eyes (Observe closely the detail).
- Respect for People.

There are 3 major flows in TPS:

- Material flow.
- Operators flow.
- Information flow.

These are designed to deliver value to the customer. Customer 1st is the major principle of Toyota Production System. The Flow of Value is most important for the customer.

After learning many good things (TPS, Shingo Model and so on) I strongly recommend that you create your own way. The Flow System builds on TPS in helping you to do that.

Please enjoy this book!

Ritsuo Shingo
President
Institute of Management Improvement

Preface

A lot has happened in just a short amount of time. We wrote this book during the summer of 2019. The book is a compilation of years of research from the fields of complexity, leadership, organizational theory, psychology, and team science. It draws on years of experience in the disciplines of engineering, military safety, and strategy throughout various organizations involved in implementing and practicing agile and lean methodologies. Since the end of August 2019, however, several events have unfolded that sparked a global pandemic from the COVID-19 virus, causing a complex environment to emerge around the globe. This new complex environment will be with us for some time. In response to the COVID-19 outbreak, a preface was warranted before this book went to press to position The Flow System within the current complex environment, in which we will be living moving forward. For context, the following events provide a review of the complex environment that had surfaced in this short time frame:

- Originating in Wuhan, China, the SARS-CoV-2 coronavirus emerged toward the end of 2019. On December 31, 2019, China alerted the World Health Organization of an outbreak due to SARS-CoV-2. This virus had then proceeded to spread globally into what is now known as the COVID-19 pandemic and had since spread to 70 other countries (Scripps Research Institute, 2020).
- Across the United States, educational institutions have shut their doors indefinitely to students. These educational institutions have transitioned to online delivery to complete this year's academic curricular obligations. The impact that this transition will have on broader educational institutions will become irreversible, as this transition could be more permanent than most may be willing to consider.
- Manufacturing facilities have come to a halt, with China shutting down toward the end of 2019 and other parts of the globe shutting down in the early

part of 2020. The COVID-19 pandemic has completely halted an entire supply chain network for manufacturing around the world. Even if China was able to begin ramping up their facilities, as they were the first hit with the virus, given the probability that they would be the first to recover, the rest of the world remains inoperable. Estimates are predicting that these supply networks will not be up and running until, at best, fall 2020. Unfortunately, most estimates dismiss the potentiality of a reoccurrence or relapse of COVID-19. Corporations are having to find different suppliers other than China; they now are looking to local manufacturers and suppliers. Reliance on China is likely to be reduced moving forward as corporations will need to have in place multiple networks in case one supplier is unable to deliver: "Redundancy is built into that diversified supplier base, which enables a quicker rebound" (n.a., 2020: Long Road). These dynamics place the global supply chain network as a complex adaptive system that will look quite different once we move beyond the COVID-19 pandemic.

- Many corporations are repurposing their facilities to provide products that are in short demand, especially those deficits found in healthcare. Some of these include Dior, who repurposed their facilities in Italy to produce hand sanitizers (*Knowledge@Wharton*, 2020). Tesla, Ford, and GM announced repurposing their facilities to provide ventilators and other medical equipment (Cormack, 2020). Honeywell, 3M, and GE are agreeing to alter facilities to produce face masks, hand sanitizers, and other shortages in hospital supplies (Smith, 2020). With the closing of manufacturing facilities, you have unemployment. Recent closures have contributed to an unprecedented unemployment rate in the United States that is "three times worse than the peak of the Great Recession" (Cox, 2020: para 9). Similar unemployment figures are being realized around the globe. These global dynamics are setting the stage for new emerging corporations for the post-COVID-19 age, producing innovative products for a different kind of customer. To survive in the future, pivoting product lines to meet current demands, even if these demands require producing a completely new product, will be the new norm expected of corporations moving forward.

- National security is at risk because of COVID-19. One such example comes from the inability of the *USS Roosevelt* to continue operating as it became grounded in Guam as a result of illness among its personnel. After identifying three members with the virus, standard procedural actions called for removing these three people to an isolated healthcare facility for treatment. Unfortunately, after a brief time, two days later, the aircraft carrier was forced to shut down indefinitely, placing our national security at risk. Although the U.S. Navy, and other military divisions, are known for being capable of responding quickly in times of emergency, the actions taken followed protocol and best practices, placing personnel in danger as a result of the COVID-19 outbreak. The protocols and best practices followed were not suitable for complex issues or problems; they were designed for reacting to known problems. For complexity, new techniques must be developed and practiced.

- The healthcare industry already had been growing with high demands for qualified personnel to meet this growth. With the current pandemic, health-care workers have placed themselves at risk, with many paying the ultimate price. China estimated that more than 3,300 healthcare providers were infected, with about 22 deaths. Similarly, Italy experienced an infection rate of approximately 20% among healthcare providers, with a few deaths (Editorial, 2020b). Because of the COVID-19 pandemic, some are claiming that the United States will soon "run out of health care workers who are essential in fighting the sprawling pandemic" (Semotiuk, 2020). Similar shortages are expected in other regions, especially Africa and Latin America, where the pandemic is just beginning to surface.

The SARS-CoV-2 virus had been identified as being a natural virus and not manufactured. Researchers determined that this virus was "the results of natural selection and not the product of genetic engineering" (Scripps Research Insti-tute, 2020: "Evidence for natural selection"). The COVID-19 virus naturally emerged in a way that was possible for the virus to change hosts and jump to humans. Similar events have occurred in recent history. For example, in 2003, the Severe Acute Respiratory Syndrome (SARS) epidemic surfaced in China, which was a virus that transferred after being exposed to civets. In 2012, the Middle East Respiratory Syndrome (MERS) emerged in Saudi Arabia, which had moved from camels (Scripps Research Institute, 2020). The agent transferring the COVID-19 virus is likely a bat, as this virus is similar to a bat coronavirus. Other potential sources could be from pangolins, armadillo-like mammals from Asia and Africa (Scripps Research Institute, 2020).

Response to the COVID-19 pandemic has resulted in countries implementing social distancing efforts; testing and isolating those infected; calling for lock-downs at local, state, and country levels; and rationing health services as a result of shortages in healthcare equipment, supplies, and providers. These protocols are in place to reduce the global death rate to only 1.9 million, an estimated sav-ings of nearly 40 million lives (*Nature*, 2020). At the time of this writing, the total number of infected people approached 500,000, with estimates of between 23,000 (*Nature*, 2020) and 39,356 (Gale, 2020) deaths globally. The spread of COVID-19 is exponential: "It took 67 days for the 100,000 cases to be reported, but just 3 days to go from 400,000 to 500,000 cases" (*Nature*, 2020: 17:35). Several experi-mental test trials are in progress, but we have no vaccine to date. Even with a vaccine, it does not necessarily protect us from any potential new threat or vari-ation that may emerge from COVID-19.

These types of viral threats will be more commonplace rather than historic anom-alies. Although the actions taken to prevent the spread were critical to avoid COVID-19 from spreading to unimaginable heights, these efforts still were reactionary measures. Reactionary measures are behaviors to manage an event after it already has occurred. In this case, the social–physical distancing, calls for lockdowns, and isolation of those infected are reactionary measures. These are best practices played out on a global scale. Unfortunately, the human race cannot afford to survive on "best practices." These are not effective strategies for protecting the human race; this could be the beginning of the end of our life cycle—the worst–case scenario.

Early signs had been ignored (e.g., SARS, MERS). This is a point that we make in The Flow System; we stress the importance of identifying and acting upon weak signals. Previous events should have raised warning signs that a pandemic was highly probable. This is evident in the following statement: "The damage wrought by SARS, Middle East respiratory syndrome, Ebola virus, Zika virus, the 2009 H1N1 influenza pandemic, and a widespread acceptance among scientists that a pandemic would one day occur" (Editorial, 2020a: 1011). Rather than countries and local communities stumbling to develop a game plan and course of action, programs already should have been made, agreed upon, and put into effect once the COVID-19 outbreak had reached the local community. For example, healthcare providers have to determine the best course of action for their personnel (e.g., doctors, nurses, staff, surgeons). This created a situation in which "most institutions do not have set policies yet, and providers at high risk levels of severe illness from COVID-19 continue to report to work" (Kofman and Hernandez-Romieu, 2020: para 8). Measures to identify, test, and develop a vaccine should have taken a priority, with any necessary resources being made available to do so. If the weak signals were acknowledged earlier on in the outbreak, the essential resources already would have been made available.

Countries and government officials should have been on the same page with the same global reaction around the world. Instead, we experienced delayed responses from various government leaders, ranging from suspended funding to support measures in the United States and other countries, to some countries ignoring the virus entirely. It wasn't too long ago, May 2018, that President Trump's administration nixed the U.S. pandemic preparedness office leaving them "flat-footed in confronting the virus" (Shesgreen, 2020: para 10). Other examples include President Obrador of Mexico (also known as AMLO), who was "dismissive and outright irresponsible" (Felbab-Brown, 2020: para 1), failing to adopt any measures to counter the spread of COVID-19. The Mexican government did not ban nonessential public events until March 24, 2020, which was counter to President Obrador's previous calls for socializing, hugging, and kissing people (Felbab-Brown, 2020). Local officials who recognized the threat were forced to counter President Obrador's directives and take measures locally to protect their communities. Other countries had similar political battles and debates on what course of action was necessary and how evasive these measures would need to be. These behaviors led to delayed responses by governments in which the citizens would be mostly affected:

> The initial slow response in countries such as the UK, the USA, and Sweden now looks increasingly poorly judged. As leaders scramble to acquire diagnostic tests, personal protective equipment, and ventilators for overwhelmed hospitals, there is a growing sense of anger. (Editorial, 2020a: 1011)

These debates also have centered on what impact these measures would have on the country's economy (e.g., gross domestic product, unemployment rate), all items irrelevant to the COVID-19 pandemic, and its eradication. In the end, these actions delayed much-needed safety measures to be implemented, resulting in the adoption of "best practices" being delayed, and making them just "practices" at best.

These points showcase the disconnect that exists between leaders (government officials) and those closest to the problem (scientists and healthcare workers). Decisions by those at the top are insular from the problem. For those dealing directly with the problem, they often must operate with a lack of resources, information, and support. These frequent examples highlight the importance of leadership; leadership must be distributed and not directional. Distributed leadership calls for those closest to the problem to be capable of making decisions and taking action to meet the demands of any threats they may encounter. Leadership is designed to support these actions by fostering enabling constraints rather than placing roadblocks in the form of inhibiting constraints, and preventing them from acting. The Flow System highlights distributed leadership as a hybrid leadership model that distributes decision-making capabilities to those closest to the problem or customer. Those at the executive levels provide support functions and strategic plans, but necessary decision making must be local.

The tools and practices used for managing or operating in complexity are not the same tools and methods required for complicated or clear problems. Best practices, a technique for clear and complicated issues, could have worked to help reduce the impact of the COVID-19 pandemic in a pre-COVID-19 era as a precautionary and preventive action. Implementing these practices in a post-COVID-19 era places these practices as reactionary and only chases the problem; these practices work for known but not complex issues. The COVID-19 outbreak created a complex environment on a global scale as soon as it became a pandemic. We are just beginning to see the ill effects of using tools and techniques designed for complicated problems in a complex environment. You cannot utilize current "best practices" to resolve complex issues; it requires complexity thinking and new techniques. For example, although current events to repurpose manufacturing facilities to make up shortages in much-needed healthcare supplies are necessary, these efforts came much too late. Too late for many. Exaptation, or repurposing existing facilities or products for a new innovative purpose, is one of the known techniques for surviving in complex environments. These efforts should have been triggered immediately with the detection of the appropriate weak signal. To move in a rapid manner is essential to halt the spread of an epidemic, preventing it from reaching the level of a global pandemic. This quick response would require community members, healthcare workers, local government officials, logistics and supply chain experts, and researchers to act together as diverse and capable teams in a multiteam system. This response would require resources and support from leadership at the global, national, and local levels, providing them with the capabilities and resources for the multiteam systems to act as they deem necessary to contain the spread of the virus.

These actions connect all of the components of the Triple Helix of Flow, integrating complexity thinking, distributed leadership, and team science into a cohesive unit. The Flow System, and this book, provide details for each of these three helixes, integrating them into a coherent system. You cannot address significant complex problems with leadership alone; it requires teams and multiteam systems. This integration is what constitutes flow, the seamless transition from ideation to delivery. The Flow System introduces a new system of understanding how to manage in disrupted, complex, and ambiguous environments. The methods, tools, and techniques are provided in this book but are also available in a free online guide book at https://flowguides.org/index. php. Currently, this guide has 10 different translations available, with more expected. Although this book, *The Flow System: The Evolution of Agile and Lean Thinking for the Age of Complexity*, presents the same methods, techniques, and tools, it also provides the research behind The Flow System and explains each of the three helixes as well as how their interconnectivity creates the Triple Helix of Flow.

These are just a few examples that demonstrate the impact caused by the COVID-19 outbreak that emerged into the pandemic in which we are now living. These are complex and ambiguous times, indeed. Although we understand that The Flow System is not the only source for necessary tools and techniques, The Flow System is an excellent starting point for many.

We do not yet know what this new complex environment will look like as we transition into a post-COVID-19 era, but we do know that we need a new kind of thinking for the post-COVID-19 age. We are living through the effects of this pandemic and must realize two essential facts. First, the global landscape will be different in the post-COVID-19 time. The world, as we know it, no longer exists, and what it looks like in the future will be up to us. Second, it is critical for nations, countries, states, communities, and organizations to learn, adapt, and begin to implement new methods, techniques, and tools that are designed for complexity. This transition can begin by utilizing the techniques provided in The Flow System. We are hopeful that The Flow System; the integration of the Triple Helix of Flow; and the methods, techniques, and tools provided in The Flow System will aid everyone during this new journey. We trust this will give us, the human race, the capabilities to prevent a pandemic from occurring a second time.

Our appreciation and gratitude go to the healthcare communities, scientists, and researchers working tirelessly to develop a vaccine for COVID-19. *Thank you!*

References

Cormack R. (2020, March 19) Tesla, GM and Ford offer to make ventilators if there's a shortage due to Covid-19. *Robb Report.*

Cox J. (March 30, 2020) Coronavirus job losses could total 47 million, unemployment rate may hit 32%, Fed estimates. *CNBC.*

Felbab-Brown V. (2020, March 30) AMLO's feeble response to COVID-19 in Mexico. Available at: https://www.brookings.edu/blog/order-from-chaos/2020/03/30/amlos-feeble-response-to-covid-19-in-mexico/.

Gale J. (2020, March 30) Dutch scientists find a novel coronavirus early-warning signal. *Bloomberg Politics.*

Kofman A and Hernandez-Romieu A. (2020, March 25) Protect older and vulnerable health care workers from Covid-19. *STAT.*

Editorial. (2020a) COVID-19: Learning from experience. *Lancet* 395(10229): 1011.

Editorial. (2020b) COVID-19: Protecting health-care workers. *Lancet* 395(10228): 922.

Knowledge@Wharton (2020, March 17) Coronavirus and supply chain disruption: What firms can learn. *Wharton Business Daily.* Available at: https://knowledge.wharton.upenn.edu/article/veeraraghavan-supply-chain/.

Nature. (2020, March 31) Coronavirus latest: lockdowns in Europe could have averted tens of thousands of deaths. *Nature.*

Scripps Research Institute. (2020, March 31) COVID-19 coronavirus epidemic has a natural origin. *Science Daily.*

Semotiuk AJ. (2020, March 31) Solving the Covid-19 crisis will require more foreign health care workers. *Forbes.*

Shesgreen D. (2020, March 18) "Gross misjudgment": Experts say Trump's decision to disband pandemic team hindered coronavirus response. *USA Today.*

Smith C. (2020, March 19) Honeywell, 3M, and GE ramp up effort to produce hospital supplies in Coronavirus fight. *Barron's.*

Endorsements

This team of highly regarded individuals recognized as experts in their fields have come together to develop the most comprehensive book for how to create flow in complex environments. The authors offer a rich toolset that is grounded in the natural sciences. The Flow System is a potential game changer for business, government, and healthcare. The authors show that this new flow system is built on a solid lean foundation starting with the Toyota Production System and Toyota Way. This system builds a new structured approach by combining three helixes of distributed leadership, complexity, and team science. This book debunks many of today's management books based on selected cases in favor of tools built on empiricism and explicit knowledge. Every leader in your company from executives to lean implementers who want to know how to develop agile companies and significantly grow their businesses should consider this a must-read.

Charles and Daniel Protzman
Partner and Vice President of Customer Solutions
Business Improvement Group LLC
danprotzman@biglean.com
www.biglean.com

If you are in the space of organizational transformation, this book is a must-read. John Turner, Nigel Thurlow, and Brian Rivera masterfully weave together an array of sciences and decades of learnings that will elevate both executives' and practitioners' depth of understanding of the critical aspects necessary to achieve excellence. They strip away superficial, window dressing, and silver bullet facades that have plagued many industries. They lay the foundation of systems thinking and principle-based architecture to a portfolio of insights on complexity thinking, team science, and distributed leadership, making this resource one of a kind in today's market. The book is easy to read, understand, and comprehend. Rather than presenting biases to frameworks, it instead sheds light on approaches organizations should consider as they go through an enterprise transformation.

Mohamed Saleh, PhD
Co-Founder & Principal
Vizibility, LLC
www.vizllc.com

Table of Contents

PART I

Introduction to Flow and The Flow System

CHAPTER 1

Introduction

O ur reality, and what should be the reality of many multinational organizations, is that our capabilities and abilities to accomplish our goals lay in manufacturing excellence. This manufacturing excellence most notably is known as the Toyota Production System (TPS). Many other companies and organizations globally have tried to replicate this system, with varying levels of success, to gain a competitive edge within their industry. Although many of these industries manufacture products other than automobiles, the TPS has become a model for organizations aiming to achieve manufacturing excellence at the highest level. The TPS is built on the pillars of *jidoka* (i.e., built with quality and having the ability to stop a process in the event of a problem) and *just in time* (i.e., having exactly what is needed only when it is needed). Built on a foundation of *standardization*, the TPS establishes a repeatable and predictable process. Among the desired goals for the TPS are the *customer*, the *employee*, and the *company*. Each desired outcome is associated with a foundational goal. The foundational goals for the customer is to place the *customer first*, that is, considering the needs and desires of the customer when determining direction and strategy. The foundational goal for the employee is *respect for humanity*, that is, targeting full creativity, development, challenging work, and trust with leadership. The final foundational goal for the company is *elimination of waste* (Muda), which is targeted at reducing overproduction, waiting, conveyance, processing, inventory, and correction (see Figure 1.1). The primary focus of the TPS, with its concept of standardization and *kaizen* (i.e., the philosophy of continuous improvement), is the customer, and now this approach has been replicated by numerous industries.

Recently, the TPS has begun to be applied to entities other than manufacturing (e.g., banks, healthcare, organizations, and universities). These applications have been extremely successful in revamping one's commitment to the customer,

3

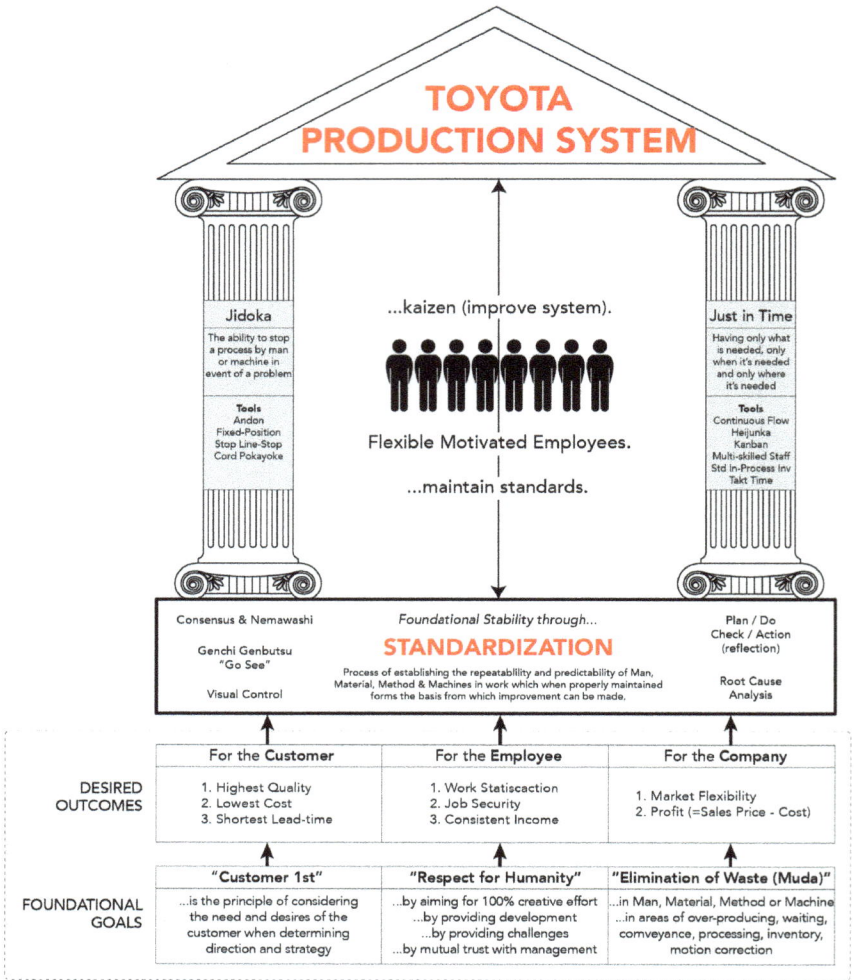

TOYOTA PRODUCTION SYSTEM

...kaizen (improve system).

Flexible Motivated Employees.

...maintain standards.

Jidoka

The ability to stop a process by man or machine in event of a problem

Tools
Andon
Fixed-Position
Stop Line-Stop
Cord Pokayoke

Just in Time

Having only what is needed, only when it's needed and only where it's needed

Tools
Continuous Flow
Heijunka
Kanban
Multi-skilled Staff
Std In-Process Inv
Takt Time

Consensus & Nemawashi

Genchi Genbutsu
"Go See"

Visual Control

Foundational Stability through...

STANDARDIZATION

Process of establishing the repeatability and predictability of Man, Material, Method & Machines in work which when properly maintained forms the basis from which improvement can be made.

Plan / Do
Check / Action
(reflection)

Root Cause
Analysis

	For the **Customer**	For the **Employee**	For the **Company**
DESIRED OUTCOMES	1. Highest Quality 2. Lowest Cost 3. Shortest Lead-time	1. Work Statiscaction 2. Job Security 3. Consistent Income	1. Market Flexibility 2. Profit (=Sales Price - Cost)
	"Customer 1st"	"Respect for Humanity"	"Elimination of Waste (Muda)"
FOUNDATIONAL GOALS	...is the principle of considering the need and desires of the customer when determining direction and strategy	...by aiming for 100% creative effort ...by providing development ...by providing challenges ...by mutual trust with management	...in Man, Material, Method or Machine ...in areas of over-producing, waiting, comveyance, processing, inventory, motion correction

FIGURE 1.1.   The Toyota Production System (TPS)

improving inefficient processes, and reducing overall waste. Although these cross-industrial implementations have been successful and probably will remain so in the future, these practices only help existing entities become more effective in delivering the same product to the customer in a more efficient manner. When the product changes or the customer base shifts, however, the whole process needs to be modified or redesigned. This system is slow to adapt to major changes.

In today's environment, with ever-changing globalization and complexity, manufacturing excellence will get you only so far. Organizations need to take an entrepreneurial approach to rapidly respond to these global changes to react quickly to new customer demands. These new customer demands apply to the product that they are seeking, and their expectations for the organizations with which they choose to do business. Customers are expecting advanced technologies to be embedded into their products and to provide interconnectivity with their existing products in a user-friendly format.

Globalization also includes a constantly evolving political environment, in which changes in one region of the globe can have negative effects across the globe, preventing new creative products from entering the market in a timely manner. When new products are introduced to market, their life spans continue to be reduced as a result of technological advances that are being developed at the same time. Innovating (i.e., the process of bringing a creative product to market) is a continuous process that places constant customer demands on the producer. New products need to have the capability to cater to each customer in real time. In its totality, this is known as operating in a complex environment.

There are, however, differences in the level of complexity one is dealing with. First, a brief distinction between complicated and complex is in order. Complicated refers to a product, system, or environment that has so many moving parts that no one person can keep up. Examples include an airplane, an electric utility power plant, or a naval vessel. To manage or service a complicated system, tools such as engineering schematics, control systems, simple feedback mechanisms, and sensors can be utilized. In contrast, complex products, systems, or environments are more than just complicated, they involve processes and characteristics that are unknown and cannot be mapped or tracked. Examples of complex systems include the brain and the universe. Although complex systems cannot be managed directly, they can be facilitated so that the desired outcomes are obtained, much in the same manner that learning leads to new behaviors in psychology (e.g., classical conditioning).

If one is dealing with low levels of complexity or with a complicated system that has little to no level of complexity, then the TPS tools and improvements provide a high degree of success for any organization or institution. When dealing with higher levels of complexity, a new framework is necessary. This is where The Flow System (TFS) comes into play. TFS is designed to enable organizations and institutions to manage and operate in complex environments. It is a framework for creativity and innovation, for tackling highly complex problems (even wicked problems[1]), and for providing the highest level of value to the customer with the quickest turnaround time possible.

---

1   Wicked problems are identified by the following characteristics: there is no definitive statement of the problem, there is no clear solution, variables are constantly changing, and affected stakeholders change over time (Roberts, 2000).

Current Examples of Complexity

Today's complexity is just beginning, and it will only continue to become more complex in the future. This is evident by the breadth of problems that organizations, institutions, and governments are facing today. The following examples highlight this reality.

Autonomous Vehicles

In the future, it is predicted that car ownership will go down exponentially. This transfer of ownership will extend to fleet-type companies that provide self-driving automobiles to the customer when demanded. Individuals will not need to purchase, maintain, and operate an automobile when they can have a car pick them up and drive them whenever they desire. Experts claim that car ownership will decline as more and more people "rely on Uber- and Lyft-type apps that deliver self-driving cars on demand" (Campanella, 2018: para. 11). Although this will provide a cheaper product for the consumer and produce value to the end user; it also will disrupt more than a few industries.

Leading experts claim that this technology, autonomous vehicles, will "disrupt or revolutionize the future of transportation as we know it" (Campanella, 2018: para. 2). Automobile insurance will change from individuals to a few fleet-type companies, requiring the number of agents to be reduced significantly as well as lowering the cost of coverage resulting from safer driving and lower accidents as a result of using self-driving automobiles. Deaths caused by automobile accidents has been among the top-ten global causes of death according to the World Health Organization, increasing in rank from 2000 to 2016 globally (World Health Organization, 2018). By reducing the human factor, we should see a significant reduction in the number of incidents, sufficient enough that the insurance industry will be disrupted.

The automobile industry also will be disrupted by having fewer customers to cater to. Car dealerships will become nearly extinct as most fleet-type companies will deal directly with the manufacturer to reduce costs. Manufacturing labor and full-time employment will be reduced as these manufacturing companies transition from fuel to electric vehicles, and then transition again from electric to autonomous vehicles. Layoffs already have been realized by some of the leading car manufacturers with estimates of autonomous vehicles hitting the market around 2025 (Swearingen, 2018). These layoffs will not be regenerated as manufacturing electric and autonomous vehicles require less labor, albeit more technically trained labor. This leads to the following statement relating to one of the larger U.S. automotive corporations: "The future of GM's profitability may lie in autonomous vehicles, but the future employment of autoworkers does not" (Swearingen, 2018: last para.). The labor market will experience a disruption as will the labor unions, nationally and internationally.

Space Exploration and Travel

Space exploration already has begun its transfer into the private sectors. Space exploration holds the potential to reduce intercontinental travel time significantly while cutting down on fuel costs and the number of emissions poured into the atmosphere each year. This Earth-to-Earth travel provides the potential to allow for city-to-city global expeditions in approximately one hour (Strauss, 2017). But such travel plans extend well beyond Earth-to-Earth travel; space exploration, hotels, and travel to Mars are just a few of the services currently being discussed for which the final destination is not Earth. Companies moving forward with space exploration are start-up companies rather than the traditional airline companies. One example is in Virgin Galactic, which expects "to be operating a variety of vehicles from multiple locations to cater for the demands of the growing space-user community" (Virgin Galactic, 2019: Spaceline for earth). A second example highlights SpaceX's plans to provide "fully reusable launch vehicles that will be the most powerful ever built, capable of carrying humans to Mars and other destinations in the solar system" (SpaceX, 2019: Advancing the future). The growth of this new exploration industry is just beginning to take off. This is illustrated by the growing number of companies getting into the business of commercial exploration. Beyond those already mentioned, other companies that recently have entered the market include Blue Origin (blueorigin.com), the United Launch Alliance (UAL; ulalaunch.com), Roscosmos (en.roscosmos.ru), and the National Aeronautics and Space Administration (NASA; nasa.gov).

Disruptions in the traditional space industry have been realized with the retirement of NASA's space shuttles. NASA has been forced to collaborate both internationally and with the private sector to continue its vision of space exploration and in sending humans into space. Airline companies and their employees will be disrupted as quicker and cheaper means of intercontinental travel are developed and expected by consumers as stewards of the environment.

Airline manufacturers along with their manufacturing subsidiaries will be disrupted significantly. Even if these industries are not already paying attention, it may not be too late. The ability to adapt to these potential disruptions in the future requires a new type of thinking from those who intend to remain in business.

Artificial Intelligence

Artificial intelligence (AI) and machine-learning technologies have just started to make advanced improvements that only now are beginning to affect our daily lives. Many of these advancements have been positive. AI is expected to exceed the capabilities of humans in most tasks within the next 10 years or so (Wilson, 2017). New technologies hold the potential to provide a higher quality of services and care at reduced costs to the customer in all areas, especially healthcare and transportation (Partnership on AI, n.d.).

A number of precautionary warnings have been made when it comes to AI and machine learning. For example, Stephen Hawking was quoted as saying, "It will bring great disruption to our economy, and in the future AI could develop a will of its own that is in conflict with ours" (Titcomb, 2016: para. 5). One strong opponent of AI technologies, Elon Musk, said, "There will be fewer and fewer jobs that a robot cannot do better" (Wilson, 2017: Elon Musk) than humans. This warning, if realized, will require humans to evolve to a future in which these AI technologies are readily available "or else become irrelevant" (Wilson, 2017: Elon Musk).

The future workplace will be involved, in one way or another, with AI and machine-learning technologies. For people looking to enter tomorrow's workplace, there is no better opportunity to find a job than if you have a background in AI technologies (Wilson, 2017). This technology adds to the previously stated disruptions in that the future workforce will be intelligence-based as opposed to labor-based. This disruption also will apply to most future jobs, which is reiterated by the Partnership on AI (www.partnershiponai.org/about/#pillar-3), an organization claiming that AI holds the potential to disrupt jobs as well as how business is conducted in the future. Although providing benefit to society and to the global economy, care must be taken to ensure that these technologies are "safe, trustworthy, and aligned with the ethics and preferences of people who are influenced by their actions" (Partnership on AI, n.d.: Safety-critical AI).

Disruptive Landscape

The examples discussed thus far are but a few of the ever-changing environmental and external forces that exist in today's landscape—this is known as operating in a complex environment. These predictions may occur differently than presented here, but they will disrupt multiple industries regardless of how then end up taking place.

Companies that are not learning to navigate and manage in this complex environment will be those that are disrupted and, in many instances, forced to go out of business. Companies capable of changing to be more adaptive to this complex environment more likely will be the companies that provide answers and new services that take over while watching competitors that are unable to adapt. These points are highlighted in the following:

> What's become clear is that no sector of the economy is safe, that the disruptions are accelerating, and that the very talented and highly trained business leaders responsible for the majority of the world's economy do not have the right set of tools and models to properly assess risk and capitalize on opportunity. (Kersten, 2018: 9)

Adaptive companies will be more capable of surviving and reformulating their products and services in times of complexity compared with those companies who try to streamline their existing products thinking only that that will provide sustainability in the future.

A Change in Thinking: The Flow System Components

Manufacturing falls under the realm of systems thinking (system approaches such as cybernetics, information theory, systems theory; von Bertalanffy, 1972) in which systems are designed to produce a prespecified product, the outcome, for example, being automobiles. This practice, however, cannot remain as the industry is beginning a new transition or phase shift. The automotive industry is at an inflection point (Kersten, 2018) in which software-based innovation is beginning to take over the electromechanical aspects of the automobile. These software-based systems add mobility and interconnectivity between each manufacturer and automobile owner as well as between automobile owners. Mixing in additional manufacturers and their automobile owners, car rental companies with their customers, and mobility companies such as Uber and Lyft with their customers, the landscape all of a sudden, and almost overnight from some perspectives, becomes an extremely open and complex environment.

Complexity Thinking

To operate in complex environments, a transition needs to take place from that of systems thinking to one of complexity thinking. Systems thinking has been successful to date in dealing with mostly closed systems, such as in manufacturing. When dealing with mostly open systems, however, this type of thinking falls short. Open systems (i.e., the environment, geopolitical landscape) have no boundaries, whereas closed systems are reduced to operating within predefined boundaries.

Complexity can be described in the following manner: "Complexity is focused on how the real-world interactions of many diverse individuals create structures (and are influenced by the structures they create) and which persist and coevolve within an environment which provides resources" (Varga, 2014: 12). Operating within the realm of complexity is dealing with connected open systems in which boundaries cannot be placed around components or agents[2] because of these interconnections.

Examples of open systems have been identified as being related to global warming, social movements, and even terrorism (Turner and Baker, 2019). Others have differentiated traditional economic theory as being simplistic, relating to systems thinking. For example, Foster (2005) identified that modern economics was formed based on the conceptualization of a simplistic system that has not been validated by historical events. This simplistic system is a result of the development of many theories in economics having been derived without

---

2   Boundaries around components refers to boundaries that typically are placed around systems, subsystems, and processes in a closed system that cannot be placed easily around open systems (i.e., social activist groups). Agents represent individuals or followers, that is, people who are taking orders or direction. This is in contrast to principals who are the delegates in a principal–agent relationship.

accurately replicating real-world conditions, similar to laboratory experiments in some disciplines. When positive results occur from these bounded experiments (laboratory experiments), and then are tested in the real world, results can be completely opposite of what was expected. The reason for this difference often is due to the influences from other environmental or external forces that were not accounted for in the original laboratory tested model. Complexity thinking, this new way of thinking, is the first component of TFS.

To operate in a complex environment, the question arises as to how one manages and operates this environment with some level of confidence on the basis of predictability. In short, shifting from systems thinking to complexity thinking is a first step. Although it is recognized that this shift is necessary, this transition will be required only when dealing with complexity. When operating in contrasting domains, such as in simple and complicated domains (Kurtz and Snowden, 2003), those that have linear cause-and-effect relationships, systems thinking will be sufficient. To address the challenge of moving organizations toward the twenty-second century, the creativity and innovative processes that evolve must be adaptive and capable of responding in real time to the demands placed on a company while operating in a complex environment.

Distributed Leadership

Implementing this new way of thinking requires an expansion of leadership at all ranks, both within an organization as well as at community and governmental levels. Distributed leadership, the second component of TFS, is critical in implementing this new way of thinking, while also being essential to managing in complexity. Leadership must be present not only as a top-down function, but also must be culturally accepted, and in some instances expected, to be driven as a bottom-up process. This systemic leadership is also horizontal in that everyone holds the potential to influence others, regardless of their hierarchical position within the organization, community, or government. Borrowing from multiple perspectives and theories of leadership, TFS's distributed leadership views leadership as an adaptive, multidimensional, shared, and engaging construct that occurs at all levels (i.e., a multilevel construct).

Before an organization can become adaptive, its members must have a leadership model that provides them with opportunities to react. This cannot come from a top-down (command and control) leadership model. Reacting to external threats requires members to think on their feet and to make decisions in real time. This requires leadership that is distributed throughout the organization with shared responsibilities and goals. Engagement not only comes from engaging the workforce but also requires leaders to be more engaged. This two-way engagement model takes place within a safe culture that provides distributed leadership among all members as opposed to a top-down leadership model that requires only employees to become engaged.

Team Science

One way to accomplish this agile and distributed leadership is by structuring activities using teams. Teams operating under a shared leadership model distribute responsibilities to all team members. When multiple teams are operating on a common goal, as in multiteam systems (MTS), this shared leadership becomes scaled up with different roles and responsibilities for the teams within the MTS. This form of teaming replicates high-performance teams that operate in complex environments around a common goal. Utilizing knowledge from the field of team science, the third component of TFS, provides us with empirical knowledge about how to structure and manage teams in complex environments.

Constraints and Barriers

Most organizations today are at a stage at which they feel the need to hold on to existing practices, processes, and structures, while also feeling the need to change to meet the demands of complexity. This feeling of holding on, while needing to change, is not new. What is new, however, to today's managers and employees is the need to learn to operate in both environments simultaneously. This duality requires operating within the domains of simple and complicated environments and in environments that provide the tools and resources needed to successfully manage in complex domains. Although the TPS is an idea model to address the former, no fully developed model addresses the latter. Hence, TFS was conceptualized based on the experiences and knowledge gained from practices implemented to deal with new product development that addresses complex problems. TFS acknowledges the need to be transformational to achieve *flow.*

Flow is described as the interactions between agents operating using complexity thinking, distributed leadership, and team science principles to become more adaptive to provide value to the customer. *Nagare* is the Japanese term for flow, which is both positive and dynamic and moves steadily and freely. From our experiences, this positive movement provides optimal value to the customer only when each of the three components of TFS are interconnected. We understand the following limitation:

- TFS is designed to address complex issues, to manage and operate in complex environments, and to develop new and innovative products and services, and it is not designed for operating in simple and complicated environments.
- TPS is the best process for operating in the simple and complicated domains and TFS is now a model for dealing in complex environments.

Constraints also are identified as being systems and human capital, not just technology. This distinction is a critical point that needs to be recognized. When operating in simple and complicated domains, systems matter, and failure of these systems often is the result of systemic breakdowns as opposed to people or technology. When operating in the complex domain, systems are open so they become nearly irrelevant, but human capital is critical in that the knowledge, skills, and interactions (KSI) of the workforce must be developed to operate in this new environment. Again, technology is not the constraint, but human capital might be if not sufficiently skilled or adequately trained, or if prevented from interacting with one another.

Given the recent advantages in technology along with the rapid decrease in the cost of acquiring these technologies, essentially every firm now has access to current technology. Competitive advantage as a result of being able to adapt to changing environmental and external demands comes not from having access to new technologies, but rather from being able to "understand and adapt the technology to meet customers' real needs" (Denning, 2018: xxvi).

Adapting new technologies in a complex environment to meet the customers' needs is achieved by being adaptive with a workforce that utilizes complexity thinking. Operating in complexity guided by TFS supports employees through the factors of complexity thinking, distributed leadership, and team science. Current managerial practices at every level must either be changed or altered to sustain in the future. These changes must focus on individual agents, teams, and their interactions, within and between teams, as opposed to focusing on processes (Denning, 2018) and best practices. TFS is designed to facilitate these interactions rather than to try to manage individuals and teams. This book reviews each of these processes in more detail and provides a synthesis of how these factors are interconnected to achieve maximum flow, that is, TFS. The following section highlights the different sections in the book along with what this book is about and what this book is not about.

Highlights of the Book

This book is organized into three parts. Part I introduces this book, highlights the importance of the customer to any organization, especially in regard to TFS, and introduces TFS and its components. Part II introduces the first component of TFS, complexity thinking; Part III presents the second component, distributed leadership; and Part IV discusses the third component, team science. In closing, Part V identifies some of the current practices and lessons learned that we have experienced in developing and in beginning the transition to adopting TFS. The final chapter concludes and provides a summary of the core values and attributes of TFS.

PART I

Part I introduces to the reader concepts of TFS and explains how TFS is designed to deliver value to the customer.

Chapter 1. Introduction to TFS

This chapter introduces TFS and the book, what TFS is and what it is not, and why it is needed at this time.

Chapter 2, Customer First, reiterates the foundation of providing value to the customer, which is the number one goal for the TPS and is the foundation for the Toyota Way, and will remain the foundation for TFS.

Chapter 3, The Flow System, introduces the components of the system along with the triple helix that interlinks each of the components into one system. Brief descriptions of Chapter 2, Customer First, and Chapter 3, The Flow System, are provided next.

Chapter 2. Customer First

Building the right product, at the right time, in the right way, while meeting the demands of the customer and consumer is, essentially, what agility is all about. In keeping with the traditions of the Toyota Way, the TPS and Lean Production, has been the "best practice" for others to follow for decades. Reducing waste that doesn't add value to the raw materials being transformed into the final product has resulted, when done correctly, to what is known today as *Lean Manufacturing* (originating with Toyota). In today's rankings among global car manufacturers, Toyota is ranked first 1 with positive growth from year to year (focus2move, 2019). The TPS, however, has been found to have some weak linkages in its application. For example, Liker (2004) has highlighted that many organizations, when adopting the TPS, tend to focus too much on the tools and not enough on the entire system. Leaders also were found not to be engaged in the daily operations (Liker, 2004). These shortcomings cannot be accepted in complex and disruptive markets. Delivering value to the customer must be the focus throughout the entire process. TFS is built on the foundation of providing value to the customer, building the right product at the right time. The purpose of business, as stated by Peter Drucker, is *"to create a customer"* (Drucker, 2007: 95). The best value for any customer in highly ambiguous and increasingly changing and demanding environments is to implement a system that flows from the customer, at the time of the initial order, to delivery of the final product. This flow can be achieved through TFS, a system designed not only to create a customer but also to keep customers in providing them the best value in complex environments.

Chapter 3. The Flow System

Chapter 3 introduces TFS and each of the components that make up the Triple Helix of Flow: complexity thinking, distributed leadership, and team science.

Agile methods and tools have been adopted from many organizations across the globe, providing tools for organizations to manage in an environment that is increasingly volatile, uncertain, complex, and ambiguous (VUCA). Unfortunately, agile methods and tools do not differentiate between complicated and complex environments or problems. Their methods and tools are one size fits all. TFS addresses the differences between clear/simple/obvious, complicated, complex, and chaotic domains (the Cynefin framework) and offers a set of methods and tools (complexity thinking) that are empirically supported to manage in complexity.

One of the biggest barriers when transitioning to becoming an agile organization is the resistance from current management structures. Moving to agile "requires a reversal of some fundamental assumptions of twentieth-century management" (Denning, 2018: xvi). This transition is, in itself, a disruptor in that it changes the structure of management (roles and responsibilities) from a traditional top-down structure to a distributed, shared, self-organizing, flatter structure that focuses on facilitating shared resources and interactions among its agents. Unfortunately, for this organizational transformation and restructuring, no agile methods or tools are available. One component of TFS is distributed leadership, which provides methods and guidance in navigating this organizational transformation.

Agile has been operating using team-based structures; teams and teams-of-teams under the framework of scrum. Another barrier to agile is in scaling—that is, how does an organization remain agile when scaling its structure to teams-of-teams-of-teams? In general, the answer is they don't. Using scrum to achieve agile has been shown to be unsuccessful on a large scale. The third component of TFS presents team science. Team science is a field of study that provides methods and tools to guide organizations in managing team-based structures, including team-of-teams, called MTS.

PART II

Part II of this manual introduces the second component of TFS in more detail, and includes Chapter 4, Systems and Complexity Theory; Chapter 5, Complexity; and Chapter 6, Systems Versus Complexity Thinking.

Chapter 4. Systems and Complexity Theory

Chapter 4 introduces both systems theory and complexity theory. Organizational transformation is discussed as this is essential for adopting team-based structures and for operating in complex environments. Making sense of unknown-unknowns and of ambiguity in complex environments is achieved through sensemaking tools and techniques, which is covered in this chapter. The Cynefin Framework is introduced with a description of the five domains: clear/simple/obvious, complicated, complex, chaotic, and disorder (Kurtz and Snowden, 2003). Following an introduction of these conceptual models, the chapter presents different tools and methods for operating in the four domains that customers may face: clear, complicated, complex, and chaotic (while also trying to avoid disorder).

Chapter 5. Complexity

Chapter 5 provides a brief introduction to what complexity is while discussing different types of complex and wicked problems along with their defining features.

Tools and techniques to address these types of problems are then introduced. This chapter provides information to identify the different types of complex problems followed by steps to address these problems.

Chapter 6. Systems Versus Complexity Thinking

Chapter 6 provides a contrast between systems thinking and complexity thinking with examples to show how systems thinking does not provide adequate tools for solving problems once the environment changes into a complex environment. Complexity thinking is presented along with guidelines about how to begin implementing complexity thinking tools to address dynamic and ambiguous environments. Complexity thinking requires acknowledgment of the domain we are operating in, and thus, the chapter presents characteristics of complex systems to help readers differentiate complicated from complex systems. Complexity thinking also requires viewing systems as complex adaptive systems. We present the characteristics of complex adaptive systems along with examples to help readers identify with these characteristics. Complexity thinking is relatively new to most fields and disciplines, and this chapter introduces this concept, along with TFS, to provide new methods for readers to utilize on an as-needed basis. These methods become most useful for readers operating in complex and ambiguous environments. This chapter presents complexity thinking to help leaders, managers, and individuals change their current thinking processes when dealing with complex problems and environments.

Part III

Part III of this manual discusses in detail the component of distributed leadership and how it interlinks the different components of TFS: complexity thinking and team science.

Chapter 7. A Word About Leadership

Before delving into leadership theories or shared, global, or instrumental leadership, this chapter discusses what leadership is and offers some words on developing leaders. Differentiating between leadership and management is covered along with the concept of leadership as a collective or team construct rather than an individual construct. This chapter describes a team leadership model designed for leadership development to highlight the concept of leadership as a team construct. Leadership capacities are introduced and described along with current trends in leadership theory.

Chapter 8. Team and Distributed Leadership

Distributed leadership incorporates a blended model of leadership that spans team-based leadership models (shared leadership) with existing hierarchical systems at the executive levels. For organizations that implement team-based structures without altering leadership style to one that is accommodating to teams often results in failed transformations. This chapter presents the concept behind team leadership, along with different models of team leadership. Next, shared leadership is introduced as one model that can be utilized for team-based structures operating in complex environments.

Chapter 9. Strategic, Instrumental, and Global Leadership

Chapter 8 presents shared leadership as a team leadership model that can be incorporated for team-based systems. As a blended model for leadership, distributed leadership incorporates shared leadership for team-based systems, with strategic, instrumental, and global leadership at the executive levels. This chapter discusses each of these leadership theories and explains how they work within the distributed leadership framework.

Part IV

Part IV introduces the third component of TFS: team science. The field of team science has grown in recent years and is a multidisciplinary field of study. The knowledge base from the field of team science provides theoretical and empirical support for managing, leading, and operating team-based systems. This book section introduces some essential elements from the field of team science that are relevant to team-based organizational structures and to TFS.

Chapter 10. Teams, Teamwork, and Taskwork

Chapter 10 covers specifics on what a team is, what a team is not (group), and what teamwork and taskwork entails, and it identifies differences between the two. One of the main shortcomings that has been identified with some agile techniques (e.g., scrum) is that they lack attention to teamwork. This chapter highlights what teamwork is and describes each component of teamwork in some detail so that team-based structures and team-based techniques can begin to incorporate teamwork into their practice.

Chapter 11. Team Effectiveness

Team effectiveness is not just a measure of a team's performance. Chapter 11 discusses what team effectiveness is and how it can be conceptualized not only to increase a team's performance but also, and more important, to provide more resilient and adaptive teams that are capable of addressing complex problems and environments.

Chapter 12. Multiteam Systems: Scaling

When scaling teams, organizations experience multiple problems and failures. Chapter 12 provides a discussion of the research on MTS and highlights some of the critical components that must be designed into any team-based structure to be successful. Tools and techniques for managing MTS also are presented.

Part V

Part V provides information about current practices that we have developed to successfully implement TFS, techniques to provide value to the customer, lessons learned, and valuable resources.

Chapter 13. Conclusion

Chapter 13 is slightly different than most concluding chapters. Remember, *flow* is contextual. Keeping that in mind, this concluding chapter does not build a case for following our model to achieve flow. What works for one organization, large or small, may not work as effectively for your organization or start-up. What is included in this concluding chapter, however, is a set of principles that synthesize the empirical evidence that composes TFS. It is our goal to aid readers in achieving flow, not by following our model, but by having a guide, TFS core principles and attributes. This chapter provides these principles and attributes for others to use as a guide for their organizational transformation and journey toward being more agile.

What This Book Is About

This book is about TFS. It includes the critical components that make up TFS, complexity thinking, distributed leadership, and team science. This book highlights each of the components by diving deep into each concept and providing evidence-based practices and knowledge gained from practice. This book provides a perspective that is rare in today's environment, one that includes both a practitioner and an academic point of view. This book closes the practice–theory gap by synthesizing the knowledge, experiences, best practices, successes, and failures of those operating in complexity. Our intention is to provide a pragmatic view of how we have conceptualized flow, that is, the ability to become more adaptive in the world of complexity.

As with many other models or frameworks (e.g., scrum, lean), context matters. What works for one organization may not work for a separate organization simply because of the contextual differences. In fact, even within the same organization, the context can change significantly. With this in mind, we acknowledge this context and understand that approaching agility in a technology startup likely will be much different from approaching agility in a healthcare setting. Because of these contextual settings, this book does not provide prescriptive frameworks; instead, it provides the critical components that we feel are required to become more agile. How one refines each of these skills (i.e., complexity thinking, distributed leadership, team science) within their environment, given their contextual setting, will vary. The information presented identifies critical components that need to be developed but does not provide prescriptive details on how to develop these skills, as this will be different for each organization.

Multiple Perspectives

This book is compiled from the knowledge, experiences, and practices that have been realized from multiple corners of today's agile environment. Input was solicited from the agile and lean communities to capture current practices and opinions. Consultants and agile and lean practitioners operating inside many multinational corporations contributed to the conversation and to many conceptual and pragmatic ideas presented in this book. Academics also contributed to the discussion and provided much-needed theoretical and evidence-based support for many of the ideas presented. This book provides the results of closing the gap between academia and practice—TFS is a product generated by academics and practitioners coming together. TFS is a theoretical system of understanding that informs theory-to-practice and practice-to-theory unification.

Acknowledgments

This book could not have been written without the full support and leadership of the co-author, Nigel Thurlow, and from his team and the management ranks at Toyota Motor North America (TMNA) and Toyota Connected (TC). As former TC Chief of Agile, Nigel provided necessary direction and support based on his experience as a practitioner operating in a world of complexity. Other consultants include Dave Snowden from Cognitive Edge (cognitive-edge.com). In conjunction with colleagues, Dave derived the Cynefin Framework as a sensemaking model for decision making. This framework has been expanded on in recent years through the efforts of Dave and his team at Cognitive Edge. The Cynefin Framework is a critical component to TFS. Other consultants include Brian (Ponch) Rivera, founder of AGLX (www.aglx.consulting) and co-author of this book. Based on Ponch's experience as a U.S. Navy pilot, he has investigated how complexity affects both business and the military. Ponch also trains busi-nesses on the Cynefin Framework as an outside consultant for Cognitive Edge. Additional training from Ponch includes training for high-performance team-ing, red teaming, Scrum the Toyota Way, Planning, and issues relating to safety incidents that occur in the military.

Academic practitioners include knowledge and expertise from Rose Baker, PhD, and John R. Turner, PhD. Rose is currently an assistant professor at the University of North Texas and contributed her expertise from the field of project management and adaptive learning. John, co-author of this book, is an assistant professor at the University of North Texas. John is also the editor-in-chief of *Performance Improvement Quarterly* (PIQ; onlinelibrary.wiley.com/journal/19378327), a Wiley academic journal, and is the developer and manager of the *Team Science: Research, Technology, and Methods* website (science-teams.com). John contributed his expertise from the field of team science, leadership, complexity, and theory building. Both John and Rose have worked together on a number of research projects surrounding teams, leadership, agile, complexity theory, systems theory, and MTS.

Charlie Protzman is the president and CEO of Business Improvement Group (BIG; biglean.com) and had consulted on this book by providing professional contextual edits and recommendations as well as including invaluable stories and historical perspectives on many of the concepts presented in the book.

Many thanks and acknowledgments go to Taz Alam for her excellent work on the graphics for this book. Taz is a freelance graphic designer (Thinktaz.com).

What This Book Is Not About

This is not a book about agile. A number of books discuss agile, agility, and a number of other synonyms. This is not one of them. This book, however, does explain how flow can be achieved from practitioners and academics working in the complexity domain. While approaching flow is the goal for TFS, it is not an agile model in the sense that TFS is conceptualized to better address complex problems based on the knowledge we already have gained. It is a pragmatic model as well, a model to help us reach the second half of the 21st century and beyond.

This is not a prescriptive book that provides the steps needed to be taken to approach agility. Although some may claim to have prescriptive formulas, we assert that achieving agility is contextual and requires different levels of emphasis on each component depending on the context, situation, available resources, and human capital, as well as on the level of complexity being addressed. Some organizations or startups may be able to develop a general prescriptive outline for their purpose, but these prescriptions are not neatly transferable to other organizations or startups. In addition, trying to scale up, or scale down, these prescriptions will lead to further complications. It is, for these reasons, that we steer away from providing a prescriptive outline for achieving agility. Instead, we concentrate on the components that we have found to be most essential for any organization or startup, in either the public or private sector, to concentrate their time and effort in developing to achieve agility. These critical components consist of the factors that lead to flow through TFS: complexity thinking, distributed leadership, and team science.

References

Campanella E. (2018, July 14) The future of self-driving cars, explained. *Global News.*
Denning S. (2018) *The age of Agile: How smart companies are transforming the way work gets done.* New York, NY: AMACOM.

Drucker PF. (2007) *People and performance: The best of Peter Drucker on management.* Boston, MA: Harvard Business School Press.

focus2move. (2019, February 21) *World cars brand ranking: The top 50 in the 2018.* Available at: https://focus2move.com/world-cars-brand-ranking/.

Foster J. (2005) From simplistic to complex systems in economics. *Cambridge Journal of Economics* 29: 873–892.

Kersten M. (2018) *Project to product.* Portland, OR: IT Revolution.

Kurtz CF and Snowden DJ. (2003) The new dynamics of strategy: Sense-making in a complex and complicated world. *IBM Systems Journal* 42: 462–483.

Liker JK. (2004) *The Toyota way: 14 management principles from the world's greatest manufacturer.* New York, NY: McGraw-Hill.

Partnership on AI. (n.d.) *Partnership on AI: About us.* Available at: https://www.partnershiponai.org/about/#pillar-3.

Roberts N. (2000) Wicked problems and network approaches to resolution. *International Public Management Review* 1: 1–19.

SpaceX. (2019) *About SpaceX: Capabilities & services.* Available at: https://www.spacex.com/about/capabilities.

Strauss N. (2017, November 15) Elon Musk: The architect of tomorrow. *Rolling Stone.*

Swearingen J. (2018, November 27) Self-driving cars are the future. Jobs in auto manufacturing are not. *New York Magazine.*

Titcomb J. (2016, October 19) Stephen Hawking says artificial intelligence could be humanity's greatest disaster. *The Telegraph.*

Turner JR and Baker R. (2019) Complexity theory: An overview with potential applications for the social sciences. *Systems* 7: 23.

Varga L. (2014) Complexity science: The integrator. In: Srathern M and McGlade J (eds) *The social face of complexity science: A festschrift for professor Peter M. Allen.* Litchfield Park, AZ: Emergent Publications, 11–25.

Virgin Galactic. (2019) *Vision: Where we're heading.* Available at: https://www.virgingalactic.com/vision/.

von Bertalanffy L. (1972) The history and status of general systems theory. *Academy of Management Journal* 15: 407–426.

Wilson C. (2017, August 16) *Artificial intelligence predictions: Peril & possibility.* Available at: https://www.iotforall.com/artificial-intelligence-predictions/.

World Health Organization (2018, May 24) *The top 10 causes of death.* Available at: https://www.who.int/en/news-room/fact-sheets/detail/the-top-10-causes-of-death.

CHAPTER 2

Customer First

The Purpose of an Organization

W hen you ask people, What is the purpose of an organization? or Why does a company exist? the most common answer you will be given is to make a profit. If, however, an organization's primary goal was simply to make a profit, it wouldn't remain in business very long. Profit is not a long-term strategy for any organization. In responding to the perception that the goal of an organization is to make a profit, Sarasohn and Protzman provided the following response:

> But such a statement is not a complete idea, nor is it a satisfactory answer because it does not clearly state the objective of the company, the principal goal that the company management is to strive for. A company's objective should be stated in a way which will not permit of any uncertainty as to its real fundamental purpose. (Sarasohn and Protzman, 1998: 1)

What is the purpose of an organization? When the primary focus is on profit, as indicative of management's fixation on quarterly results, then the organization's purpose has replaced the customer with profit. In many instances, the customer has been completely displaced from the organization's focus. Further insight on this topic is provided again from Sarasohn and Protzman:

> It is entirely selfish and one-sided. It ignores entirely the sociologic aspects that should be a part of a company's thinking. The business enterprise must be founded upon a sense of responsibility to the public and to its employees. Service to its customers, the wellbeing of its employees, good citizenship in the communities in which it operates—these are cardinal principles fundamental to any business. They provide the platform upon which a profitable company is built. (Sarasohn and Protzman, 1998: 1)

These ideas are not germane to management theory today nor are these organizational philosophies widely taught in today's business schools. These ideas and philosophies are not new. The statements from Sarasohn and Protzman came from the postwar era as they were part of the rebuilding effort in Japan led by General Douglas MacArthur (Sarasohn and Protzman, 1998). This rebuilding effort included the Civil Communications Section (CCS) that were tasked with rebuilding the communications systems in Japan. At the time, the communications systems in Japan were in complete shambles. Also, from this rebuilding effort, with direct ties to Sarasohn and Protzman, came Edward Deming. Deming is well known for his work with Toyota in the early years, but initially he was involved with contributing to the rebuilding efforts in Japan after the war.

It's no wonder, then, after reading the perspectives provided by Sarasohn and Protzman, that Deming was heavily influenced and had similar beliefs: "The consumer is the most important part of the production line" (Deming, 2000: 26). Deming not only believed in knowing what the customer wanted but also emphasized the importance of knowing what the customer thought about the product after experiencing the deliverable: "'GOOD QUALITY' and 'UNIFORM QUALITY' have no meaning except with reference to the consumer's need" (Deming, 1952: 5). Additional connections and similarities were found between Deming and Peter Drucker. In describing the purpose of a business, Drucker stated: "There is only one valid definition of business purpose: to create a customer" (Drucker, 2007: 95). In his essential principles for managers, Drucker included as one of these principles.

> "The single most important thing to remember about any enterprise is that there are no results inside its walls. The results of a business is a satisfied customer. Inside an enterprise, there are only cost centers. Results exist only on the outside" (Drucker, 2006: 196).

Never Skip on Quality

Never skip on quality. Why? The answer is quite simple, because the customer doesn't want an inferior or unreliable product or service. So, why do so many organizations sacrifice on quality? Today, some managers still hold the belief that was prominent more than 70 years ago: that you can't have both quality and performance. This either/or philosophy believes that by pushing for performance or increases in production, quality suffers and, in contrast, pushing to improve quality is detrimental to performance or production numbers. This is the general belief that a balance must exist between quality and quantity. From this perspective, it is the customer who suffers. Managers typically choose quantity over quality, mainly because of their focus on short-term profits. Completely disregarding the customer, "defects and faults that get into the hands of the customer lose the market" (Deming, 2000: 3).

This quality or quantity balance is essentially what allowed Japanese manufacturers to gain a competitive advantage early on through the rebuilding efforts orchestrated by General MacArthur. Through Deming's teachings, and his work with the Union of Japanese Scientists and Engineers (JUSE), the Japanese manufacturers believed that improving quality also improved productivity (Deming, 2000). In contrast, Western industry believed in a balance: "to improve quality to a level where visible figures may shed doubt about the economic benefit of further improvement" (Deming, 2000: 2).

The Japanese manufacturers developed the philosophy that "improvement of quality begets naturally and inevitably improvement of productivity" (Deming, 2000: 2). Support of this philosophy came from Walter A. Shewart's book *Economic Control of Quality of Manufactured Product*, in which he showed that reductions in variation, through quality measures, results in improvements in production, quantity (Deming, 2000). The model for the Japanese was described as being a chain reaction that later became known in Japan as a way of life (Deming, 2000). According to Deming (2000: 3), the steps for this way of life, or culture, include the following:

- Improve quality.
- Decrease costs because of less rework, fewer mistakes, fewer delays, and fewer snags, and better use of machine-time and materials.
- Improve productivity.
- Capture the market with better quality and lower price.
- Stay in business.
- Provide jobs and more jobs.

It's Part of the Culture

Moving forward to today's market and organizational philosophies, focusing on quality is still front and center in delivering value to the customer. This philosophy today is the same as when it was presented by Deming, JUSE, and others in the 1950s: "The customer is the most important part of the production line. Quality should be aimed at the needs of the customer, present and future" (Deming, 2000: 5).

In looking at how Toyota's culture has evolved over the years, not much has changed. The culture of continuous improvement—that is, improving quality and processes—refers to improving value to the customer and not just improving within. The Toyota Way involves the two components of continuous improvement and respect for people. The continuous improvement includes increasing quality by reducing waste to provide increasing value to the customer. The respect for people involves not only respecting one's employees but also fostering an unfolding respect for the customer, suppliers, stakeholders, associates, business partners, and the community at both the local and global levels. This responsiveness to the community comes from Toyota's focus on long-term prosperity for all, not just within. This long-term focus includes being a good corporate citizen; providing employment to support local regions and countries; and developing a productive, educated, and responsible workforce. Development involves producing a workforce that can self-reliantly put the Toyota Way into practice. Only then, can value be provided to the customer.

Customer focus is evident throughout Toyota. One such example can be found in the values for Toyota Motor Sales (TMS), which are similar to the Toyota Promise and Lexus Covenant:

- We must never stand still. Where others might rest, Toyota will move forward and seek out the opportunity to do even better.
- We honor our customers as welcome guests and serve them in the manner they desire.
- We respect the time and promise of our customers and colleagues. (Liker and Hoseus, 2008: 498)

The Toyota Production System (TPS) adds value to the customer by reducing non-value-added waste (Liker, 2004). The customer is front and center throughout the TPS, as evident in the following examples that highlight how TPS begins with the customer: *"The need for fast, flexible processes that give customers what they want, when they want it, at the highest quality and affordable cost"* (Liker, 2004: 8; emphasis in original). *"Because the only thing that adds value in any type of process—be it in manufacturing, marketing, or a development process—is the physical or information transformation of that product, service, or activity into something the customer wants"* (Liker, 2004: 9–10; emphasis in original).

These points highlight the culture in delivering not only the best product to the customer, whether this product is a physical delivery or a service, but also the right product to the customer at the right time. Building the right product at the right time, at the level of quality expected by the customer, and at the right cost has become the norm in today's economy. Meeting such demands, when the customer is demanding more variety in products and services, creates a complex environment. This new level of complexity is only expected to increase as the industry becomes disrupted with growing demands for electric and auto-driving cars from a new type of customer (e.g., Uber, Lyft). This growing level of complexity is just beginning to be realized and is expected to increase as the industry becomes increasingly disruptive in the coming years. To meet complex demands better and faster and to adapt to flexible customer demands is the focus of The Flow System (TFS).

Why Change Is Needed Now

During the development of the TPS, The Toyota Way, and Toyota's culture of Customer First, a few things have changed. First, Deming originally believed that the customer was not the one who produces a product, rather customers only know what they want or need after being introduced to a new product or service by the producer. Deming expressed this concept in the following: "The customer expects only what the producer has led him to expect" (Deming, 1994: 8). This mind-set was later reinforced in other works from Deming: "New product and new types of service are generated, not by asking the consumer, but by knowledge, imagination, innovation, risk, trial and error on the part of the producer" (Deming, 2000: 182).

Although this point may seem simplistic, it still resonates. For example, consumers did not know that they wanted or needed a television until they first saw a television and there was a network to provide regular news and entertainment. Before Apple Computers began their journey to provide computers for the average consumer and not just for businesses, people never really thought about owning a personal computer. Once, however, the product was delivered to the consumer, everyone grew to expect a personal computer and today it is commonplace rather than a rarity.

Drucker did not disregard the consumer, quite the opposite. The consumer was well educated on the product or service that was being offered and became a critical component in evaluating the utility of any product or service. This is how consumer research began to take hold.

A second essential point that was driven by Deming's work was that a producer must understand the customers' experience with their product. This feedback process uncovers "how his [the producer] product performs in service, what people think of his product, why some people will buy it, why others will not, or will not buy it again, and he is able to redesign his product, to make it better" (Deming, 2000: 178).

The initial problem with feedback from the consumer is that first impressions are emotional, not based on experience or facts. It is only when consumers have had enough time using the product that they can provide factual feedback based on personal experiences with the product. The former feedback is a biased indicator and can be useful to measure consumers' reactions to a product design or other aesthetic characteristics. The latter is evidence-based feedback from the consumers' experiences. The more useful indicator of a product's usefulness and quality comes after the customer has experienced the product. This second measure is a lag indicator, meaning that some time has elapsed between producing a product and receiving feedback on the usefulness of that product. Deming stated this point in the following manner:

> The customer's reaction to what he calls good service or poor service is usually immediate, whereas reaction to the quality of manufactured product may be retarded. How a customer will rate a product or service a year hence or two years hence can not be ascertained today. Judgment of the consumer may shift with respect to service as well as with manufactured product. His needs may change. Alternate choices of service may appear on the market, as with manufactured product. Moreover, service may deteriorate. Manufactured product may be subject to latent defects. (Deming, 2000: 186)

Today, the customer has changed from these initial behaviors as the environment has changed. Customers not only are more educated about what they expect from the companies with which they do business but also are more vocal about what products and services they expect. This changes the feedback from the customer to *fast feedback*, which, at times, is unsolicited. Fast feedback comes with some advantages, for example, it allows smaller queues in production and has the potential to provide unlimited information surrounding each phase or prototype in the process (Reinertsen, 2009). This fast feedback requires companies to learn and adapt their products and services to meet current customer expectations. This quick responsiveness expected of the producer has changed their role from being passive to being aggressively responsive.

Organizations are developing continuous connections to better meet the demands of today's customer. Identified as a *seismic shift*, corporations and institutions are beginning to change: "Instead of waiting for customers to come to them, firms are addressing customers' needs the moment they arise—and sometimes even earlier" (Siggelkow and Terwiesch, 2019: 66). Companies must move beyond their traditional model, known as "the age of buy what we have" (Siggelkow and Terwiesch, 2019: 73), or risk being disrupted by new and more responsive organizations.

A Trust Issue

Evidence of companies unable to attend to the new demands of the customer is available in the news every day. Consumers are beginning to lose trust in the organizations they do business with. Some of these examples include Facebook in 2018, when they admitted that hackers accessed log-in information of nearly 50 million users, or when they gave access to user profiles to Netflix, Spotify, Microsoft, Yahoo, and Amazon (Sucher and Gupta, 2019).

Summary

Companies can no longer wait a year, or longer, to receive customer feedback on their product. In many cases, after a year, new products already will be on the shelves and unsatisfied customers will change brands and products. Attending to customer demands and complaints, in real time, becomes a necessity its today's environment. As this environment continues to change, with the customer making demands of producers, a new system or transformation is in order. Continuing to do business using organizational structures and theories from the past will not prove to be very successful for most companies and institutions. This changes how customer value is delivered and is the foundation of TFS, to provide value to customers in this, the new age of the consumer, and in tomorrow's complex environment.

The Flow System: Customer Value

Building on the history, knowledge, and experiences gained from Toyota, and from all those who contributed to the TPS and The Toyota Way, TFS maintains its focus on providing value to the customer. This is evident in the diagram of TFS (see Figure 2.1) in that its foundation includes the TPS and The Toyota Way. This focus is also apparent in the overarching theme of *Customer First Value Delivery* along the top of the helixes (complexity thinking, distributed leadership, team science). The next chapter, Chapter 3, The Flow System, provides a brief summary of the system of understanding and subsequent chapters provide more details. The primary focus, however, is to provide increasing value to the customer in tomorrow's disrupted and complex environments.

As Toyota's value of *Customer First* is achieved through daily practice of The Toyota Way (continuous improvement, respect for people), the authors believe that providing value to the customer in times of disruption and in complex environments require different tools and techniques than those provided in the TPS. While practicing The Toyota Way daily remains essential, TFS was conceptualized because the landscape was continuously changing and being challenged by disruptors. TFS provides the best value to the customer by identifying the means

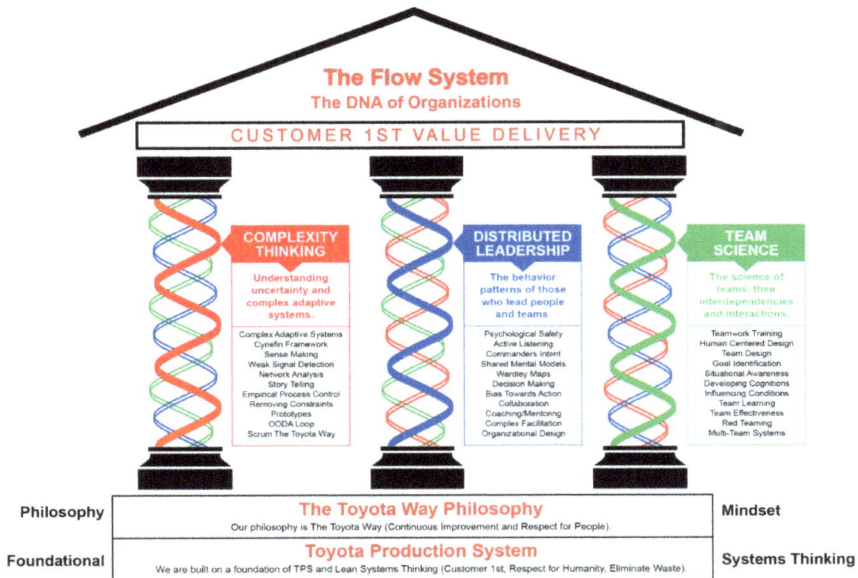

FIGURE 2.1. The Flow System

of producing FLOW[1] through the interconnections of complexity thinking, distributed leadership, and team science. Only when FLOW is achieved, by connecting all three components of TFS, can customer value be maximized. When these three components are not interconnected, bottlenecks are formed that contribute to loss of information, knowledge transfer, or incorrect information exchanges, resulting in defects and delays. These bottlenecks hinder customer value rather than increase customer value. TFS provides maximum value to the customer in times of complexity and ambiguity.

References

Deming EW. (1952) *Elementary principles of the statistical control of quality: A series of lectures.* Tokyo, Japan: Nippon Kagaku Gijutsu Remmei.

Deming EW. (1994) *The new economics: For industry, government, education.* Cambridge, MA: MIT Press.

Deming EW. (2000) *Out of the crisis.* Cambridge, MA: MIT Press.

Drucker PF. (2006) *Classic Drucker.* Boston, MA: Harvard Business School Press.

Drucker PF. (2007) *People and performance: The best of Peter Drucker on management.* Boston, MA: Harvard Business School Press.

Liker JK. (2004) *The Toyota Way: 14 management principles from the world's greatest manufacturer.* New York, NY: McGraw-Hill.

Liker JK and Hoseus M. (2008) *Toyota Culture: The heart and soul of the Toyota Way.* New York, NY: McGraw-Hill.

Reinertsen DG. (2009) *The principles of product development flow: Second generation lean product development.* Redondo Beach, CA: Celeritas Publishing.

Sarasohn HM and Protzman CA. (1998) *The fundamentals of industrial management: The Homer Satasohn papers, 1936-2001.* Series 1: Civil communications section (CSS), 1946–1998.

Siggelkow N and Terwiesch C. (2019) The age of continuous connection. *Harvard Business Review* 97: 64–73.

Sucher SJ and Gupta S. (2019 July) The trust crisis: Facebook, Boeing, and too many other firms are losing the public's faith. Can they regain it? (Reprint BG1904). *Harvard Business Review* 97. Retrieved on HRB.org

---

1   The concept of *flow* will be described in detail in the next chapter. Briefly, *flow* represents the feeling of being engaged, when nothing else matters because everything works as expected. The Japanese term for flow is *nagare*, moving steadily and freely.

CHAPTER 3

The Flow System

The Flow System

The concept of flow involves one's experience of being totally engaged, the sense of "joy, creativity, the process of total involvement with life" (Csikszentmihalyi, 1990: xi). Flow, from a psychological perspective, relates to achieving happiness by having control of one's life (Csikszentmihalyi, 1990). Within the context of the theory of optimal experience (or happiness), flow has been described as follows: "The state in which people are so involved in an activity that nothing else seems to matter; the experience itself is so enjoyable that people will do it even at great cost, for the sheer sake of doing it" (Csikszentmihalyi, 1990: 4).

From a personal viewpoint, achieving happiness effortlessly would be the best explanation of flow. One example of this would be when an artist finally reaches the stage at which they feel their product is done. Many artists struggle with finishing a product, constantly wanting to tweak something or start fresh just because they don't, or didn't feel, that the product achieved what was anticipated. Once these struggles escape the artist, and the artist feels that the piece of work is finished, a sense of relief is felt along with the joy of contributing a new piece of work.

This same feeling of joy and happiness has been found in studies looking at dance and listening to music (Bernardi et al., 2018). These studies have defined flow as "a strong rewarding experience of deep absorption and energized, focused attention" (Bernardi et al., 2018: 416). Other experiments include spoken-word artists performing in conjunction with dancers in an improvisational performance. The spoke work artist would speak while the dancer reacts to the spoken words. This improvisational performance known as *Flow!* has taken root in the United Kingdom in which two art forms, spoken word artists and dancers, come together as an *integrated whole* to achieve a type of *consciousness awareness* (Connell and Newland, 2017). In this context, flow is best described as "a narrative of in-the-moment decision-making of judgements, directions and predilections that inform the dancer's movements" (Connell and Newland, 2017: 264).

Flow is characterized as being associated with the "upswing, upwelling of life" (Smith and Lloyd, 2019: 3), as a product of "our actions and interactions with others" (Smith and Lloyd, 2019: 3). This description extends the activity of flow from being just an individual construct to a social construct. This upswing and upwelling of life occurs from the interactions with others, when structure and practice become unnoticed and all that is conscious is the act of doing.

This move to flow being a social construct implants knowledge gained from fields, such as anthropology, ecology, physics,[1] psychology, and team science, to name but a few. Flow is a result, in part, of collective motion in which individuals, or agents, learn to react to their environment to obtain their goals. Agents achieving flow through collective motion is identified in the following: "Entities that interact with their environment via explicitly modeled perceptions and actions, endowed with an internal mechanism for deciding how to respond, and capable of adapting those responses based on an individual history of interactions and feedback" (Ried et al., 2019: 2).

In The Flow System (TFS), providing customer value cannot be achieved without first interconnecting the concepts of complexity thinking, distributed leadership, and team science as an integrated organizational structure. TFS is contextual in that it is primarily conceptualized for complex, ambiguous, rapidly changing, disruptive environments rather than for dealing with the status quo. The concept of flow is an evolving process, as the components of complexity thinking, distributed leadership, and team science become more interconnected over time, flow becomes even more seamless, natural, and unnoticed. Flow as an evolving concept comes from constructal law, which states that everything that moves is a flow system and the system's configuration must evolve in such a way that provides easier access to the currents of flow through it (Kosner, 2012). TFS is one such design for organizations to evolve while navigating, successfully, disruptive and complex environments. The following sections provides a brief description of TFS, while the remainder of this book provides details of each component in the system.

The Triple Helix of Flow

TFS is composed of three essential concepts or schools of thought: complexity thinking, distributed leadership, and team science. Each of these components

---

1   From the field of physics, constructal law states: "For a finite-size system to persist in time (to live), it must evolve in such a way that it provides easier access to the imposed (global) currents that flow through it" (Reis, 2006: 269). Constructal law highlights that a system to does not develop on its own: "system shape and internal flow architecture do not develop by chance, but result from the permanent struggle for better performance and therefore must evolve in time" (Reis, 2006: 269).

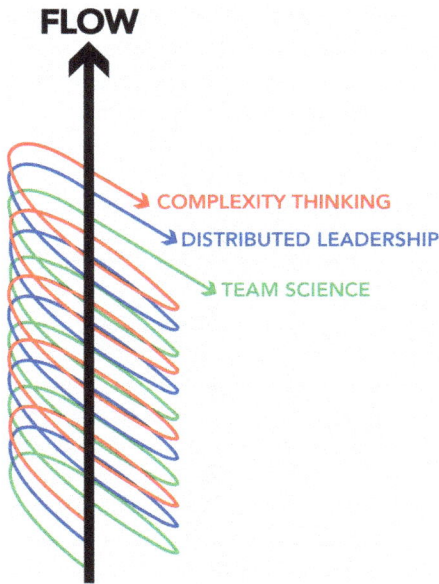

FLOW

COMPLEXITY THINKING
DISTRIBUTED LEADERSHIP
TEAM SCIENCE

FIGURE 3.1.   The Triple Helix of Flow

must be interconnected, synchronized, and embedded in an organizational structure before being able to achieve flow in delivering value to the customer in complex and disrupted environments. This integration of these three concepts is what we termed the Triple Helix of Flow and is shown in Figure 3.1.

The concept of a triple helix represents the spaces in which the interactions between and among agents emerge to become more adaptive and agile. This triple helix represents the interconnected links between complexity thinking (agility), distributed leadership, and team science principles. This new agile triple helix can be described as being similar to other conceptual models. For example, when identifying the processes of innovation, a triple helix of innovation was represented by Leydesdorff and Etzkowitz (1998). Their representation identified the triple helix of innovation that included the interactions among independent institutions (industry, government, and academia). The Triple Helix of Flow is similar with the exception of the integrating agents; here, we are looking at the patterns that emerge between and among agents through the frameworks of complexity thinking (agility), distributed leadership, and team science principles. The Triple Helix of Flow identifies the nonlinear[2] interactions that emerge into new patterns, networks, and knowledge that advances innovation theory

---

2   Nonlinearity occurs when an input does not always produce a predetermined outcome.

and practice (Triple Helix Research Group, 2019). The emergence[3] that occurs from these nonlinear interactions from the concepts of complexity thinking, distributed leadership, and team science results in an organization's capability to adapt to external forces while providing value to the customer.

A System of Understanding

TFS is presented as a system of understanding. We define TFS much in the same manner that Deming defined a system for his *System of Profound Knowledge* (Deming, 1994). Deming's system of profound knowledge involved managerial transformation through an understanding of four components: appreciation for a system, knowledge about variation, theory of knowledge, and psychology (Deming, 1994). Deming stated that managers were not required to be an expert in any one of these components, all that was required for transformation to occur is for managers to have an understanding in which they can apply the components (Deming, 1994). TFS is such a system of understanding in that no one employee, manager, or executive is expected to be an expert in all three of the components (complexity thinking, distributed leadership, team science), nor are they expected to be a master in any one of the components. What is being proposed, however, is that employees, managers, and executives have a level of understanding of each of the three components so that they can be put into practice. The benefits will be an organization that is capable of adapting to environmental variation[4] to meet the demands of the customer in today's complex environment.

TFS also can be considered an informal theory. Formal theories are those that have been tested in different contextual settings and have been accepted by a discipline as an explanation or a prediction of a phenomenon of interest to that discipline. An informal theory is a new theory that has not been tested, or has not been fully tested, preventing the theory from being adopted by any single discipline. The theoretical life cycle, or the scientific process, however, involves the introduction of alternative theories that, at times, replace formal theories for a discipline. TFS identifies the constructs of the system and within the body of this book, the interconnections between each of these three constructs will be presented. This identifies how each of the three constructs must work together to achieve *flow*. TFS could be identified as a theoretical model per the following definition of a theory: "A conceptual framework that identifies the connections, or lack of connections, between concepts/constructs to describe a phenomenon that furthers the academic knowledge base and supports researchers and practitioners in the field in which the phenomenon takes place" (Turner et al., 2018: 38).

---

3 Emergence occurs when new outcomes not previously expected occur.
4 Environmental variation refers to the frequency in which environmental conditions change (e.g., globalization, global warming).

Not a Framework or a Taxonomy

Although presented as a system of understanding and a theory, TFS is not to be mistaken for a framework or taxonomy.

A framework is a presentation of systems or concepts that are expected to be present across all, or many, domains (Snowden, 2012). It is a representation of a structure or a system that consists of categories that account for a phenomenon; frameworks describe phenomenon by categorizing items or events, but they do not provide explanations for phenomenon (Nilsen, 2015). TFS is not a framework and does not provide any type of categorization.

The purpose of a taxonomy, typically represented as a 2 x 2 matrix, is to "classify units of study by creating superordinate categories that are similar on a number of different underlying dimensions" (Hollenbeck et al., 2012: 83). Taxonomy is another tool used to categorize items or events related to a specific phenomenon. Again, TFS is not taxonomy.

A Brief Description of The Flow System

As a system of understanding, TFS highlights the essential components that lead to *flow*. Flow is constantly moving steadily and freely, like a river, with temporary boundaries and constraints. Also, like a river, the force over time will change the boundaries and restraints, causing the river to flow in the same manner but along a different path. The river will continue to flow just the same. When dealing with human social systems (e.g., team, organization, government), the boundaries and constraints are human-derived (human-made and self-constructed). A social system can either be derailed by these self-derived boundaries and constraints, or the social system can transform in such a manner that the flow changes the boundaries and restraints allowing the social system to forge ahead. TFS is a system that provides the necessary components for a social system to *flow* in times of uncertainty, complexity, and ambiguity to provide the ultimate objective for any organization, to provide value to the customer.

The components of TFS (complexity thinking, distributed leadership, team science) must be interconnected into one fluid river rather than separate smaller streams. The tools and techniques listed under each of these three components in Figure 2.1 include only a few tools and techniques that have been identified to address complexity. There are more tools and techniques than those listed, and there are more being developed as time moves on. Our intention was not to list all of the tools and techniques in one diagram; we listed only some of the main tools and techniques that we have experienced to date. Other tools and techniques will be identified and tested over time, and this list will more than likely be updated over time. This work to identify additional tools and technique is part of a larger project within our research team. Updates to the list of tools and techniques will be provided on our webpage (www.theflowsystem.com) along with a TFS Guide (flowguides.org/) that reflects these changes. This chapter provides a summary of TFS components and serves as the first guide to TFS.

This guide will be divided into three sections: complexity thinking, distributed leadership, and team science. Each of these sections will provide a brief description of the different components, tools, and techniques. Detailed descriptions, along with the empirical research findings to support the components, are provided in the remaining chapters in this book.

Complexity Thinking

The first helix that makes up the triple helix is complexity thinking and is illustrated in Figure 2.1 as follows: understanding uncertainty and complex adaptive systems (CAS). The tools and techniques for complexity thinking include CASs, Cynefin framework, sensemaking, weak signal detection, network analysis, storytelling and narratives, empirical process control, constraint management, prototypes, the observe-orient-decide-act (OODA) loop, and *Scrum: The Toyota Way*.

Complexity Thinking

When addressing complex environments,[5] the following questions need to be asked:

- What state is the current environment at?
- How much variability is in the current environment?

When the environment is a complex environment[6] that includes high variability and uncertainty, one must utilize tools and techniques that are designed for complex environments rather than for complicated or simple environments and problems. Although the goal of complexity thinking is to move complex problems[7] into the complicated domain, where we already have proven tools and techniques for addressing complicated problems, we still need a different set of tools and techniques to begin this transition phase.

Complexity thinking involves two steps:

- understanding the characteristics of complex systems; and
- having a worldview or perspective that systems, entities, and events are CASs.[8]

---

5   Complex can apply to one's surrounding or environment or it can apply to specific types of problems.

6   Complex environments include those with multiple possible states, that change rapidly, and that are unpredictable. Complex environments can already exist, or they can be self-generated.

7   Complex problems have no consensus on how to solve the problem.

8   Complex adaptive systems.

These two steps provide guidance in beginning to address complex problems and to navigate uncertain environments. Utilizing the tools and techniques provided will aid in this transition phase.

Complex Adaptive Systems

CASs typically are described as being self-organizing open systems; they are continuously dynamic, learning, and evolving to external changes (environmental forces). CASs operate between order and disorder, making them nonpredictive and nonlinear systems. Characteristics of CAS include the following: path dependency, system history, nonlinearity, emergence, irreducibility, adaptability, operation between order and chaos, and self-organization. Complete descriptions of each of the eight CAS characteristics are given in Chapter 6.

Cynefin Framework

Dave Snowden's Cynefin[9] framework identifies five distinct domains of knowledge: clear/obvious/simple, complicated, complex, chaotic, and disorder (Cognitive Edge, 2010; Kurtz and Snowden, 2003). The Cynefin framework identifies five knowledge domains that require different tools and techniques to manage. Through identification of these knowledge domains, agents are able to respond to their *sense of place* to generate new perspectives and narratives (Browning and Boudes, 2005). TFS is designed to manage and navigate the waters of complexity, providing tools and techniques for leaders to move from the complex domain into the complicated and more manageable domain.

Sensemaking

Sensemaking is about organizing "unknowable, unpredictable streaming of experience in search of answers to the question, 'what's the story?'" (Weick, 2009: 132), or what is the problem? Sensemaking leads to finding answers to the question "now what?" and is answered through ongoing updates, retrospect extracted cues, and identity plausibility (Weick, 2009). Alternatively, sense-making (with a hyphen) is similar but is derived from a different field of study then sensemaking. Sensemaking was derived from Karl Weick with a background in social psychology, whereas sense-making was derived from Dave Snowden from industry looking at complexity (Browning and Boudes, 2005). Sensemaking (Weick, 2009) displays the perception of an all-encompassing term, whereas sense-making (Snowden, 2012) represents a set of processes (Browning and Boudes, 2005). Both are similar in their descriptions with "considerable overlap on the substance

WTF

---

9   Cynefin (pronounced kuh-nev-in) is a Welch term meaning habitat (noun), and acquainted or familiar (adjective). Cynefin refers to one's environment or place of comfort in the Cynefin framework (Browning and Boudes, 2005).

of their thinking" (Browning and Boudes, 2005: 33). Both are techniques to aid our understanding of complex situations. For our purpose, we use the term sensemaking as an all-encompassing term but recognize that distinctive differences exist between sensemaking and sense-making.

Sensemaking requires narratives, or storytelling, for agents to conceptualize their surroundings and to form their own set of mental models. In the following example, many of the characteristics identified under the complexity thinking heading are also touched upon (e.g., flow, unknowable, stories, and emergence).

To focus on sensemaking is to portray organizing as the experience of being thrown into an ongoing, unknowable, unpredictable streaming of experience in search of answers to the question, "what's the story?" Plausible stories animate and gain their validity from subsequent activity. The language of sensemaking captures the realities from agency, **flow**, equivocality, transience, reaccomplishment, unfolding, and emergence, realities that are often obscured by the language of variables, nouns, quantities, and structures. (Weick, 2009: 132; emphasis added)

Weak Signal Detection

Before one can avoid failure, the pure chance that failure is possible must be realized. Embracing failure means two things, it is essential to identify weak signals and understand that mistakes are part of the learning process and should be accepted rather than punished:

> First, it means that they [high-reliability organizations] pay close attention to weak signals of failure that may be symptoms of larger problems within the system. Second, it means that the strategies adopted by HROs often spell out mistakes that people don't dare make. (Weick and Sutcliffe, 2007: 46)

Weak signal detection, also termed *increasing insights* and *noticing more*, is a capability that helps identify opportunities and threats that are present in complex environments. First coined by Igor Ansoff, weak signal detection in TFS is the ongoing process of scanning the environment from diverse perspectives to identify discrete changes to avoid surprise and capture innovative solutions.

Network Analysis

When talking about open social systems, or CASs, we essentially are talking about networks. Examples of networks include culture, nature, brains, organisms, economics, and ecologies (Borgatti et al., 2018). Networks provide a way of thinking about open social systems that "focus our attention on the relationships among the entities that make up the system" (Borgatti et al., 2018: 2). Network analysis, and other similar tools, provide a means of analyzing networks and the relationships within these networks.

Storytelling and Narratives

Narratives are a form of communication that "happens in conversation, is composed of discourse, appears in a sequence, and is interpreted retrospectively" (Browning and Boudes, 2005: 32). Narratives aid in uncovering solutions to complex and ambiguous situations through language, metaphors, and stories (Browning and Boudes, 2005).

Empirical Process Control

According to Polanyi and Prosch, "there are many variations of the advisability of being puzzled. But, in any case, a scientific explanation must serve to dispel puzzlement" (1975: 53). Dispelling this puzzlement occurs from empiricism rather than from subjectivity or dogmatic[10] practices. Science is achieved through three general ideas:

1. *Empiricism*: The only source of real knowledge about the world is experience (Godfrey-Smith, 2003: 8)
2. *Mathematics and Science*: What makes science different from other kinds of investigation, and especially successful, is its attempt to understand the natural world using mathematical tools (Godfrey-Smith, 2003: 10)
3. *Social Structure and Science*: What makes science different from other kinds of investigation, and especially successful, is its unique social structure. (Godfrey-Smith, 2003: 12)

Science is a social process that requires analysis of knowledge gained from experience to explain or describe a phenomenon. This process is often hampered when subjectivity takes control over empirical processes, introducing human biases into the process. This introduction of human bias is explained by the following sets of blindness that need to be avoided:

a) We focus on preselected segments of the seen and generalize from it to the unseen: the error of confirmation.
b) We fool ourselves with stories that cater to our Platonic thirst for distinct patterns: the narrative fallacy.
c) We behave as if the Black Swan[11] does not exist: human nature is not programmed for Black Swans.
d) What we see is not necessarily all that is there. History hides Black Swans from us and gives us a mistaken idea about the odds of these events: this is the distortion of silent evidence.
e) We "tunnel": that is, we focus on a few well-defined sources of uncertainty, on too specific a list of Black Swans (at the expense of the others that do not easily come to mind). (Taleb, 2010: 50)

---

10 Dogmatic relates to one believing that their opinion (subjectivity) is the law or truth.

11 Black Swan refers to identifying a rare or unexpected event. It is possible to identify Black Swans when avoiding biases, using empiricism, and utilizing tools that can detect weak signals.

Constraint Management

Constraints are portrayed as being constructed, as Weick notes: "Constraints are partly of one's own making and not simply objects to which one reacts" (2009: 147). This leads to the need of externalizing each agents' understanding of what constraints are present, which contributes to weak signal detection. It is not necessarily that we need to remove, reduce, or manage these constraints as much as it is to identify these perceived constraints.

As an organizational theory that recommends focusing on the weakest link, the theory of constraints presents itself as a systems-based theory. Also, the theory of constraint acts more in-line with the continuous improvement literature (Simsit et al., 2014), reverting more toward the TPS and systems thinking. Later iterations of this theory incorporated thinking processes that included cause-effect tools (Simsit et al., 2014), keeping this theory as a systems-based theory. Unfortunately, these thinking processes and tools were classified as being non-user-friendly (Watson et al., 2007). This theory is described as follows: "As can be seen it actually focuses on continuous system improvement by dealing with constraints" (Simsit et al., 2014: 934).

Given that we want to identify constraints, physical and cognitive (perceived), because we are dealing with social systems, this can be accomplished through the tools that we have identified in TFS: storytelling, mental maps, narratives, and weak signal detection. TFS recommends extending the theory of constraints for the complex domain by utilizing the tools available in TFS. Identifying constraints in complex environments is essential for weak signal detection, sensemaking, and storytelling, as well as for developing mental models. The tools provided in TFS help identify constraints in complex and ambiguous environments.

Prototypes

A prototype is a "concrete representation of part or all of an interactive system" (Elverum and Welo, 2016: 3007). Prototypes have been identified as one technique for dealing with complex problems.

OODA Loop

The OODA loop is a nonlinear decision-making process for dynamic situations. Considered to be an individual and organizational learning and adaption process, the OODA loop was developed by Air Force Col. John Boyd. Col. Boyd developed the OODA loop more than 40 years ago by blending his experience in the cockpit with science, strategy, warfare, early complexity thinking, and lessons from the Toyota Production System (TPS).

The OODA loop is a continuous cycle that can begin at either of the four components (observe-orient-decide-act). The OODA loop has been broken down into four cycles to match each of the four domains in the Cynefin framework, identifying a continuous process for each knowledge domain.

Scrum: The Toyota Way

Scrum: The Toyota Way is a training program that aids development of human capital through problem identification and definition, customer identification and acknowledgment, teamwork skill development, planning and estimation techniques, and visualizing work processes.

Distributed Leadership

The second helix that makes up the triple helix is distributed leadership (see Figure 2.1) and is described as the behavior patterns of those who lead people and teams. Distributed leadership is best described as leadership that extends horizontally, vertically, and every place in between. Distributed leadership requires shared and team leadership models while providing top-down and bottom-up processes that cross organizational boundaries and break down existing siloed structures. Our model of distributed leadership includes strategic, instrumental, and global leadership theories at the executive levels that have been shown to be supportive of team-based organizational structures and complex environments. The tools and techniques identified for distributed leadership include psychological safety, active listening, leader's intent, shared mental models, bias toward action, collaboration, coaching and mentoring, complex facilitation, and organizational design.

Psychological Safety

Among the factors listed under cooperation, perhaps the most essential factor, based on recent research findings, is to develop a high level of psychological safety among team members. As the construct of psychological safety has gained prominence in the literature, some confusion surrounding the construct persists. A few historical distinctions must be provided to dispel some of these confusions to further clarify what psychological safety is and how it can contribute to a team's success.

A distinction can be made between psychological safety (individual[12]) and team psychological safety (team, group, collective). Psychological safety as an individual construct relates to one's feeling of being "secure and capable of changing" (Edmondson, 1999: 354). In contrast, team psychological safety is a shared belief in which the team is safe from interpersonal risk taking (Edmondson, 1999). Ultimately, team psychological safety emerges as a group property (Edmondson, 2019) that is dynamic and can vary from one team to the next. As highlighted in the book *The Fearless Organization*, Edmondson summarized the

---

12 Psychological safety, as an individual construct, was first presented by Schein and Bennis in 1965 to represent "the need to create psychological safety for individuals if they are to feel secure and capable of changing" (Edmondson, 1999: 354). See also Schein and Bennis (1965).

following: "Leaders of teams, departments, branches, or other groups within companies play an important role in shaping psychological safety" (2019: 22).

Active Listening

Leadership is as much about listening as it is about talking. Active listening is just one part of a leader's ability to successfully practice effective storytelling. Storytelling is a reciprocal process of listening and telling where: "telling shapes listening" and "listening shapes telling" (Nossel, 2018: 33). Storytelling, through active listening and telling, aids leaders and followers in their sensemaking endeavors.

Leader's Intent

Leader's intent, similar to *commander's intent*, describes what constitutes success for a strategy, product, or other desired organizational outcome. In a complicated environment (known-knowns), for example, in which cause-and-effect relationships are present, the objective is to focus on the desired end-state through decentralized planning and execution of activities from those closest to the work. For complex environments (unknown-unknowns), in contrast, in which there is no direct cause-and-effect relationships, desired outcomes are unknown and are formalized through sensemaking activities through autonomous agents closest to the problem.

A commander's intent is accomplished through seven planning process steps:

Step 1. Planning Initiation
Step 2. Mission Analysis
Step 3. Course of Action Development
Step 4. Course of Action Analysis and Wargaming
Step 5. Course of Action Comparison
Step 6. Course of Action Approval
Step 7. Plan or Order Development (Joint Chiefs of Staff, 2018: II-6)

A commander's intent provides initial structure before an operation but allows those closest to the work, autonomous agents, to make their own decisions as to how to accomplish the initial goal: "It provides focus to the staff and helps subordinate and supporting commanders act to achieve the commander's objectives without further orders once the operation begins, even when the operation does not unfold as planned" (Joint Chiefs of Staff, 2018: II-7). A commander's intent aids in developing subordinates' situational awareness and decision-making capabilities in complex environments.

For TFS, the leader's intent is used because it is more recognizable to the larger organizational audience.

Shared Mental Models

Shared mental models are representative of all team members' knowledge structures relating to tasks, resources, equipment, working relationships, and situations (Turner et al., 2014; Van den Bossche et al., 2011). Building shared mental models among team members is essential to team learning in which team members' develop, modify, and reinforce mental models through interactions (Mohammed and Dumville, 2001). Team members' knowledge structures support team learning and sensemaking behaviors, contributing to more effective team decision making and team effectiveness outcomes (Van den Bossche et al., 2011).

Wardley Maps

Strategy in organizations often ignores the concept of landscape, referring to a description of one's landscape including stakeholders and constraints (Wardley, 2016). Recognizing that most "models are wrong but some are useful" (Wardley, 2016: para. 1), Wardley concentrated his efforts on developing a map-type representation of one's landscape. Wardley maps depict one's landscape and aid in developing one's situational awareness of their environment. Wardley maps create a shared mental model of a business strategy, product, or service to inform strategy and decision-making capabilities.

Decision Making

The underlying basis of the intellective-judgmental continuum (team decision-making processes) requires the following:

> (a) agreement on a conceptual system [problem], (b) sufficient information, (c) sufficient knowledge of the system [problem] by incorrect members to recognize correct solutions, and (d) sufficient ability, motivation, and time for the correct group members to present the solution to the incorrect members. (Laughlin, 2011: 141)

Decision making in groups is beneficial compared with decision-making practices used by highly skilled individuals (Hackman, 2011), making it even more important to practice essential leadership functions:

> A reasonable chance for success also requires careful attention to specifying team purposes, to selecting the right team members. To establishing the norms of conduct that will guide team behavior, and to providing the organizational and leadership supports the team will need. (Hackman, 2011: 66)

Bias Toward Action

Bias toward action is a leadership development technique in which leaders and teams learn to act on their own decisions in times of uncertainty. A bias toward action helps remove the constraints of centralized decision making by bringing the decision-making capabilities to those closest to the uncertainty, thus increasing the flow that produces maximal value to the customer.

Collaboration

Corporations and scientific institutions have begun transitioning into a process of collaboration. This transformation has been called a *team science revolution* or a new *collaboration cosmopolitanism* (Bozeman and Youtie, 2017). Collaboration cosmopolitanism is defined as "a form of interaction among producers of knowledge, allowing effective communication and exchange; sharing of skills, competencies and resources; working, generating and reporting findings together" (Ynalvez and Shrum, 2011: 205). Collaboration is an essential skill that needs to be practiced by all members, especially in open work environments and in multiteam systems (MTS; team of teams).

Coaching and Mentoring

Before a team can perform exceptionally, then someone in a leadership position needs to pay attention to the team (Hackman, 2011). Through coaching, teams can receive the attention required when needed and also may be free to operate autonomously when comfortable. The function of coaching is not to command teams, but to "help the team increase its capability to competently manage its own processes" (Hackman, 2011: 135).

Complex Facilitation

Complex facilitation techniques are complexity-based approaches that often are viewed as being counterintuitive. Complexity removes the facilitator from the content to provide for the development of shared mental models. Complex facilitation techniques in TFS utilize the Cynefin framework in conjunction with techniques, such as *ritual dissent* (alternative method of strategic planning) and *anthro-simulation* (human-machine manipulation of the environment).

Organizational Design

Organizations are beginning to realize the benefit of transforming into flatter organizations, reducing the level of hierarchies within a single organization to give more decision-making power to those closest to the work. Unfortunately, organizations are too top heavy and need to go through an organizational transformation before being able to restructure to be adaptable for complex environments and before being able to compete against disruptive competitors. Organizations typically are structured around the organization's communication structure (Conway, 1968), this has been termed Conway's law. In reality, however, we need an anti-thesis of Conway's law. Organizations need to restructure to better support team-based structures as opposed to team-based structures that are designed around an organization's communication or accounting structure. The goal is for organizations to "support the ability of teams to get their work done—from design through to deployment—without requiring high-bandwidth communication between teams" (Forsgren et al., 2018: 63).

Team Science

Team science is the third helix in the triple-helix model. Team science is illustrated in Figure 2.1 as the science of teams, their interdependencies, and their interactions. Team science is a multidisciplinary field of study that touches on the collaborative functioning of teams and small groups in the workplace, often involving cross-disciplinary and cross-functional groups. The tools and techniques identified for team science include teamwork training, human-centered design, team design, goal identification, situational awareness, developing cognitions, influencing conditions, team learning, team effectiveness, red teaming, and MTS.

Teamwork Training

Teamwork initiates interaction and the free flow of information, providing shared resources and knowledge that can be applied to any one task, thus providing a team with more ability when compared with each individual team member's skills. Team members need to be trained to manage team working skills before they can operate as a high-performing team.

Human-Centered Design

Also known as design thinking, human-centered design is a human-centered innovation process involving "observation, collaboration, fast learning, visualization and quick prototyping, and concurrent business analysis" (Lockwood, 2010: 5). Design thinking involves participation from all stakeholders, including the customer, designer, manufacturer, and suppliers.

Team Design

The type of team and the composition of team members is a critical component to any team-based organization. Teams need to be composed of members who have the skills expected to complete the team's tasks and also have an equal distribution of power and other diversity characteristics. Utilizing the right type of team and designing the team accordingly is a critical skill that managers need to learn when dealing with complexity.

Goal Identification

Teams must be able to attend to the team's goal while keeping to the organization's goal. Conflicts often arise when teams lose sight of the organizational goal and focus too much attention on the team's goals. Coordinating activities across multiple teams, as in MTS, requires an emphasis on the team's goals (proximal goals) and the MTS goals (distal goals).

Situational Awareness

Collecting, processing, and making sense of information about the contextual setting or situation relates to situational awareness. Situational awareness involves ones perception of, synthesis of, and projection of knowledge before acting or to aid in one's action (Almeida et al., 2019).

Developing Cognitions

The level of shared understanding between team members is essential for everyone to coordinate activities in achieving the team's goals. Each team member needs to be aware of who has which skills, experiences, and knowledge (transactive memory system) to perform the team's tasks. Team members must have a similar understanding of what is expected of the team and each member also must have an accurate understanding (shared mental model). Team cognition is a team-level construct that involves both individual-level characteristics (e.g., knowledge, identity, learning) as well as team-level characteristics (e.g., collective knowledge, team identity, team learning). Teams must be able to develop these team cognitions before being able to function as an effective team. Team cognitions also allow teams to learn and adapt as a collective unit as required in highly changing environments.

Influencing Conditions

Influencing conditions involve the conditions that team members have little to no control over. These influencing conditions include context, composition, and culture. Identifying how to operate under a variety of influencing conditions will aid team members in successfully achieving the team's goals.

Team Learning

Team learning occurs as a shared outcome of team member interactions. Team learning represents "an ongoing process of reflection and action, through which teams acquire, share, combine, and apply knowledge" (Mathieu et al., 2008: 431). Team learning aids in a team's ability to be adaptive and to be innovative and also leads to higher levels of task performance and efficiency (Mathieu et al., 2008). Team learning involves both the processes and outcomes of team member interactions.

Team Effectiveness

Team effectiveness can be broken down into four components: teamwork (interpersonal dynamics of team members), taskwork (technical knowledge, skills, and abilities to complete a team's tasks), the tangible outputs or products of the team (productivity, efficiency, quality, quantity; Mathieu et al., 2019), and the value delivered to the customer (including team member satisfaction; Turner,

Baker, Ali, and Thurlow, 2020). When any components are ignored, the team's overall effectiveness suffers. Managing and leading team-based organizational structures requires knowledge of all four components.

Red Teaming

Red teaming is a cognitive approach that aids in developing pathways to better decision making (University of Foreign Military and Cultural Studies, n.d.). This method uses the following set of structured tools and techniques to make better decisions:

> Help us ask better questions, challenge explicit and implicit assumptions, expose information we might otherwise have missed, and develop alternatives we might not have realized exists. It cultivates mental agility to allow Red Teamers to rapidly shift between multiple perspectives to develop a fuller appreciation of complex situations and environments. This leads to improved understanding, more options generated by everyone (regardless of rank or position), better decisions, and a level of protection from the unseen biases and tendencies inherent in all of us. (University of Foreign Military and Cultural Studies, n.d.: 3)

Multiteam Systems

MTS involve two or more component teams working toward a common MTS goal (Shuffler et al., 2015; Zaccaro et al., 2012). MTSs are similar to team of teams, the scaling of teams, and networks of teams. The structure of MTS is different from the structure of individual teams and managing each requires different managerial techniques. Managing or leading MTS requires knowledge of how MTS are composed, how the teams are linked to each other, and how the MTS develops or evolves over time.

Conclusion

To be adaptive in times of environmental change and uncertainty, corporations that are better able to push their decision-making capabilities downward are those corporations that remain successful and solvent in the long term. One example of this in practice is with Toyota's implementation of the *andon* system in manufacturing—any employee has the ability to stop production if a problem has been identified. This simple quality control mechanism showcases Toyota's commitment to moving decision-making capabilities to those closest to the work. Employees are empowered to make their own judgments in determining whether a problem does indeed exist, which extends beyond manufacturing as it is part of Toyota's culture. *Ji Kotei-Kanketsu* (JKK) stands for developing, maintaining, and continuously improving optimal work processes through collaborative and cooperative methods to continuously produce the best outputs. *Ji* means *for oneself*, that is, placing the decision making and responsibility to the lower levels of the organization.

In times of complexity, as during an economic recession or a pandemic, for example, organizational structures must be designed to move decision-making processes to those closest to the work. This allows organizations to rapidly adapt to changing conditions (Frick, 2019; Pisano, 2019). Companies are beginning to realize: "to survive in today's volatile, uncertain, complex, and ambiguous environment, they need leadership skills and organizational capabilities different from those that helped them succeed in the past" (Moldoveanu and Narayandas, 2019: 42).

Leaders also are expected to view organizations as CASs. Leaders must reject closed systems because of their inability to adapt (Saelinger, 2018). This point was driven home by General McChrystal in his analogy of forcing square pegs into round holes: "You cannot force a square peg into a round hole, and you cannot force the complex to conform to rules meant for the merely complicated" (McChrystal et al., 2015: 65).

More and more organizations today are beginning to benefit from utilizing teams and collaborative efforts. Teams have become the basic building blocks of today's organizations to meet tomorrow's "hyper competitive and fluid environment" (Mathieu et al., 2019: 18). Teams provide benefits for organizations to: "innovate, stay relevant, and solve problems that seem unsolvable" (Kwan, 2019: 68). When teamwork is disregarded and not developed, however, teams can have negative impacts. For example, medical error within the United States is the third leading cause of deaths, with teamwork failures accounting for upward of 75% of those errors that were classified as being serious medical errors (Mayo and Woolley, 2016). Although teamwork comes with many benefits and provides advantages in managing complex and ambiguous environments, poor teamwork can have negative consequences. Place the importance of team member training and developing teamwork skills before attempting to operate in any complex environment.

The military has been forced to move away from large command structures to utilizing smaller team-based forces that are capable of going from door to door. This changing military force utilized teams to better adapt to the changing environment, and previous planning and organizing methodologies no longer result in the same precision in today's unexpected landscapes (Porkolab and Zweibelson, 2018).

What is common to all of these examples is that each identify some form of complexity, leadership, and teaming, separately, with a few examples using perhaps two of these concepts. It is not revelatory to highlight that complexity thinking, distributed leadership, and team science, as presented with TFS, are crucial elements in addressing uncertainty and disruptive environments. Plenty of examples in current news and in research literature highlight these components separately. One such example is healthcare: teams have become the organizational structure because of hypercompetitive environments, but they have also resulted in poor healthcare results when teamwork training and development are dismissed. What is revelatory in TFS is that these components must connect before achieving a state of *flow*, as presented in the triple-helix concept. This is evident with the healthcare example, in which teamwork is essential to achieving a high level of patient safety and satisfaction, and also must be interconnected with leadership and complexity thinking.

The literature supports our point. There are countless books on leadership, complexity, and teams separately. A few do combine some elements of leadership and teams, or leadership and complexity, but few combine all three elements. No other book takes the position that all three components must be interconnected to achieve *flow* to deliver the best value to the customer. This is what sets TFS apart from existing literature and practice and should be the very reason why TFS will have utility across a variety of contextual settings.

[handwritten: Not convinced in TFS]
[handwritten: like the grasp of all TKS]
[handwritten: Team! Can their]

References

[handwritten: predict fall, la, capture]

Almeida RB, Junes VRC, Machado RdS, et al. (2019) A distributed event-driven architectural model based on situational awareness applied on internet of things. *Information and Software Technology* 111: 144–158.

Bernardi N, Bellemare-Pepin A, and Peretz I. (2018) Dancing to "groovy" music enhances the experience of flow. *Annals of New York Academy of Sciences* 1423: 415–426.

Borgatti SG, Everett MG, and Johnson JC. (2018) *Analyzing social networks*. Thousand Oaks, CA: Sage.

Bozeman B and Youtie J. (2017) *The strength in numbers: The new science of team science.* Princeton, NJ: Princeton University Press.

Browning L and Boudes T. (2005) The use of narrative to understand and respond to complexity: A comparative analysis of the Cynefin and Weickian models. *Emergence: Complexity and Organization* 7: 32–39.

Cognitive Edge. (2010, August 24) *Summary article on origins of Cynefin.* Available at: https://cognitive-edge.com/articles/summary-article-on-cynefin-origins/.

Connell G and Newland I. (2017) In and out of *Flow!* improvisatory decision-making in dance and spoken word. *Choreographic Practices* 8: 259–277.

Conway ME. (1968) How do committees invent? *Datamation* 14: 28–31.

Csikszentmihalyi M. (1990) *Flow: The psychology of optimal experience.* New York, NY: Harper Collins.

Deming EW. (1994) *The new economics: For industry, government, education.* Cambridge, MA: MIT Press.

Edmondson A. (1999) Psychological safety and learning behavior in work teams. *Administrative Science Quarterly* 44: 350–383.

Edmondson AC. (2019) *the fearless organization: Creating psychological safety in the workplace for learning, innovation, and growth.* Hoboken, NJ: Wiley.

Elverum C and Welo T. (2016) Leveraging prototypes to generate value in the concept-to-production process: A qualitative study of the automotive industry. *International Journal of Production Research* 54: 3006–3018.

Forsgren N, Humble J, and Kim G. (2018) *Accelerate: Building and scaling high performing technology organizations.* Portland, OR: IT Revolution.

Frick W. (2019) How to survive a recession and thrive afterward: A research roundup. *Harvard Business Review* 97: 98–105.

Godfrey-Smith P. (2003) *Theory and reality: An introduction to the philosophy of science.* Chicago, IL: University of Chicago Press.

Hackman RJ. (2011) *Collaborative intelligence: Using teams to solve hard problems.* San Francisco, CA: Berrett-Koehler.

Hollenbeck JR, Beersma B, and Schouten ME. (2012) Beyond team types and taxonomies: A dimensional scaling conceptualization for team description. *Academy of Management Review* 37: 82–106.

Joint Chiefs of Staff (2018, October 22) *Joint Operations: Joint publication 3-0.* Available at: https://www.jcs.mil/Portals/36/Documents/Doctrine/pubs/jp3_0ch1.pdf?ver=2018-11-27-160457-910.

Kosner AW. (2012, February 29) There's a new law in physics and it changes everything. *Forbes.* Available at: https://www.forbes.com/sites/anthonykosner/2012/02/29/theres-a-new-law-in-physics-and-it-changes-everything/#415dcf24618d.

Kurtz CF and Snowden DJ. (2003) The new dynamics of strategy: Sense-making in a complex and complicated world. *IBM Systems Journal* 42: 462–483.

Kwan LB. (2019) The collaboration blind spot: Too many managers ignore the greatest threat in launching cross-group initiatives: provoking defensive behaviors. *Harvard Business Review* 97: 67–73.

Laughlin PR. (2011) *Group problem solving.* Princeton, NJ: Princeton University Press.

Leydesdorff L and Etzkowitz H. (1998) Triple helix of innovation: introduction. *Science and Public Policy* 25: 358–364.

Lockwood T. (2010) The bridge between design and business [editorial]. *Design Management Review* 21: 5.

Mathieu JE, Gallagher PT, Domingo MA, et al. (2019) Embracing complexity: Reviewing the past decade of team effectiveness research. *Annual Review of Organizational Psychology and Organizational Behavior* 6: 17–46.

Mathieu JE, Maynard TM, Rapp T, et al. (2008) Team effectiveness 1997–2007: A review of recent advancements and a glimpse into the future. *Journal of Management* 34: 410–476.

Mayo AT and Woolley AW. (2016, September) Teamwork in health care: Maximizing collective intelligence via inclusive collaboration and open communication. *AMA Journal of Ethics* 18: 933–940.

McChrystal S, Collins T, Silverman D, et al. (2015) *Team of teams: New rules of engagement for a complex world.* New York, NY: Penguin.

Mohammed S and Dumville BC. (2001) Team mental models in a team knowledge framework: expaning theory and measurement across disciplinary boundaries. *Journal of Organiztional Behavior* 22: 89–106.

Moldoveanu M and Narayandas D. (2019) The future of leadership development: Gaps in traditional executive education are creating room for approaches that are more tailored and democratic. *Harvard Business Review* 97: 40–48.

Nilsen P. (2015) Making sense of implementation theories, models, and frameworks. *Implementation Science* 10: 13.

Nossel M. (2018) *Powered by storytelling: Excavate, craft, and presnt stories to transform business communication.* New York, NY: McGraw-Hill.

Pisano GP. (2019) The hard truth about innovative cultures. *Harvard Business Review* 97: 62–71.

Polanyi M and Prosch H. (1975) *Meaning*. Chicago, IL: University of Chicago Press.

Porkolab I and Zweibelson B. (2018) Designing a NATO that thinks differently for 21st century complex challenges. *Applied Social Sciences* DR 2018/1: 196–212.

Reis HA. (2006) Constructal theory: From engineering to physics, and how flow systems develop shape and structure. *Applied Mechanics Reviews* 59: 269–282.

Ried K, Muller T and Briegel HJ. (2019) Modelling collective motion based on the principle of agency: General framework and the case of marching locusts. *PLoS ONE* 14: e0212044.

Saelinger D. (2018) The end of bureaucracy: How a Chinese appliance maker is reinventing management for the digital age. *Harvard Business Review* 96: 51–59.

Schein EH and Bennis W. (1965) *Personal and organizational change via group methtods*. New York, NY: Wiley.

Shuffler ML, Jimenez–Rodriguez M, and Kramer WS. (2015) The science of multiteam systems: A review and future research agenda. *Small Group Research* 46: 659–699.

Simsit ZT, Gunay NS, and FVayvay O. (2014) Theory of constraints: A literature review. *Procedia* 150: 930–936.

Smith SJ and Lloyd RJ. (2019) Life phenomenology and relational flow. *Qualitative Inquiry*. Advance online publication: https://doi.org/10.1177/1077800419829792.

Snowden DJ. (2012, July 25) *The origins of Cynefin: Part 5*. Cognitive Edge. Available at: http://old.cognitive-edge.com/wp-content/uploads/2010/08/The-Origins-of-Cynefin-Cognitive-Edge.pdf.

Taleb NN. (2010) *The black swan: The impact of the highly improbable*. New York, NY: Random House.

Triple Helix Research Group. (2019) *The triple helix concept*. Available at: https://triplehelix.stanford.edu/3helix_concept.

Turner JR, Baker R, Ali Z, and Thurlow N. (2020) A new multiteam system (MTS) effectiveness model. *Systems* 8: 21.

Turner JR, Baker R, and Kellner F. (2018) Theoretical literature review: Tracing the life-cycle of a theory and its verified and falsified statements. *Human Resource Development Review* 17: 34–61.

Turner JR, Chen Q, and Danks S. (2014) Team shared cognitive constructs: A meta-analysis exploring the effects of shared cognitive constructs on team performance. *Performance Improvement Quarterly* 27: 83–117.

University of Foreign Military and Cultural Studies. (n.d.) *The red team handbook: The Army's guide to making better decisions*. Ft. Leavenworth, KS: University of Foreign Military and Cultural Studies.

Van den Bossche P, Gijselaers W, Segers M, et al. (2011) Team learning: building shared mental models. *Instructional Science* 39: 283–301.

Wardley S. (2016, August 8) *On being lost*. Warldeymaps. Available at: https://medium.com/wardleymaps/on-being-lost-2ef5f05eb1ec.

Watson KJ, Blackstone JH and Gardiner SC. (2007) The evolution of a management philosophy: The theory of constraints. *Journal of Operations Management* 25: 387–402.

Weick KE. (2009) *Making sense of the organization: The impermanent organization*. West Sussex, UK: Wiley.

Weick KE and Sutcliffe KM. (2007) *Managing the unexpected: Resilient performance in an age of Uncertainty.* San Francisco, CA: Jossey-Bass.

Ynalvez MA and Shrum WM. (2011) Professional networks, scietific collaboration, and publication productivity in resource-constrained research institutions in a developing country. *Research Policy* 40: 204–216.

Zaccaro SJ, Marks MA, and Dechurch LA. (2012) Multiteam systems: An introduction. In: Zaccaro SJ, Marks MA, and Dechurch LA (eds) *Multiteam systems: Am organization form for dynamic and complex environments.* New York, NY: Routledge, 3–32.

PART II

Complexity Thinking

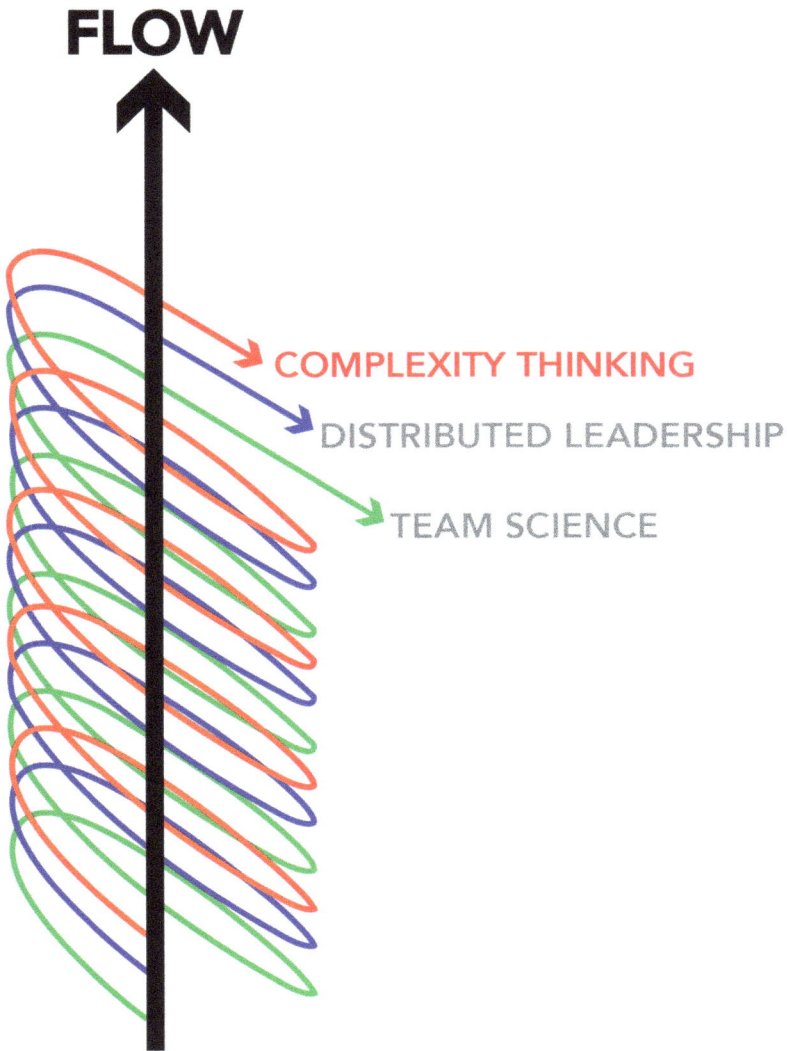

FLOW

COMPLEXITY THINKING

DISTRIBUTED LEADERSHIP

TEAM SCIENCE

CHAPTER 4

Systems and Complexity Theory

Introduction

Complexity can be found frequently in literature, books, magazine articles, and personal blogs. The topic of complexity, or volatility, uncertainty, complexity, and ambiguity (VUCA) from the management literature, has evolved rapidly to the point at which one might think that researchers and practitioners discovered a breakthrough in organizing and managing. Complexity, however, has been talked about for some time.

For example, in discussing how practitioners operate in action, Schon (1983: 39) highlighted the importance of understanding the phenomena of "complexity, uncertainty, instability, uniqueness, and value-conflict." Complexity, along with a number of the other phenomena listed earlier, present new issues, problems, and situations that haven't been experienced before by most researchers or practitioners. In these circumstances the problem must be constructed from data and information (evidence) that is "puzzling, troubling, and uncertain" (Schon, 1983: 40). This practice of constructing the problem is what Schon (1983: 40) referred to as making sense "of an uncertain situation that initially makes no sense." Complex problems[1] also add a new level of difficulty to problem solving in that the variables often change frequently. This means that traditional problem-solving techniques are ineffective at solving complex problems because of the constantly

---

1 A distinction can be made between complex and wicked problems. The distinction is recognized and is covered in the section "Complex Problems." Complex problems are those in which the problem is agreed upon, but consensus is not reached on how to solve the problem. In contrast, wicked problems are those in which there is no agreement on the problem or the solution. For more information, see Roberts (2000).

changing variables (e.g., barriers, environment, expectations, inputs). Thus, as complexity has been around for some time, techniques about how to address complex problems and how to manage in complex environments are just beginning to surface. The Flow System (TFS) is one such technique.

Transforming an organization or institution by utilizing TFS tools likely require many to change current practices, which may involve altering the culture of the organization. Transformational and cultural change will be required to transition to a system that flows.

This chapter will distinguish the characteristics of systems theory with those of complexity theory. Following is a discussion on complexity, organization transformation, and sensemaking. The chapter then introduces the Cynefin framework (Cognitive Edge, 2010; Kurtz and Snowden, 2003) as a sensemaking framework that aids in identifying which techniques, tools, and practices should be utilized based on the environment or type of problem that is being addressed. This chapter closes with a discussion differentiating between systems thinking and complexity thinking, along with a conclusion that combines the previously discussed items to create a comprehensive model of complexity thinking.

Systems Theory and Complexity Theory

Systems Theory

When talking about systems theory, one must first understand the concept of wholeness. Systems theory is an extension of the *system holistic principle* from Aristotle and the Gestalt movement, in which the whole is greater than the sum of its parts. An example includes an organization (the whole) that is made up of individuals (agents, parts). A definition for systems theory is as follows: "A theoretical framework by which elements that act in concert to produce some result are studied" (Yawson, 2012: 56). Using systems theory, the whole (the system) is broken down into parts (systems), and the parts can then be broken down even further into parts-of-parts (subsystems). By extending the previous example, an organization (the system) can be broken down into divisions (systems), and each division can be broken down further into departments (subsystems). This division allows one to concentrate on optimizing one of the systems, or any of its subsystems, to improve the overall operation of the system. Another example could be a hospital. The emergency room is a part of a bigger system, as are the laboratory and radiology departments. Each of these components, or subsystems (emergency room, laboratory, radiology), are part of a bigger system called the hospital. The overall hospital is a complex system with some of its subsystems being complex (e.g., surgery) and others being simple (e.g., hospital registration).

When breaking down each system, or any of the subsystems, systems theory utilizes the input-process-outcome (IPO) model. The IPO model breaks down each system and each subsystem into its own set of inputs, processes, and outcomes. The outcome of one subsystem potentially affects the input of other subsystems, and sometimes the overall system. Feedback mechanisms are in play throughout these systems to make them more intelligent.

To identify the general laws that are common to all systems, Ludwig von Bertalanffy (von Bertalanffy, 1968) derived what he termed general system theory (GST). GST presents common laws found for most systems: "There are general aspects, correspondences, and isomorphisms common to 'systems'" (von Bertalanffy, 1972: 415). The main distinction between the original version of GST and today's version is that GST primarily was applied to closed systems. Since the initial version, there have been many iterations of GST. GST recently has been presented most commonly as systems theory (ST). In general, ST provides a means of following the transformation of inputs into outputs by identifying a systems' inputs, processes, outputs, and feedback mechanisms (Turner and Baker, 2019). Later, von Bertalanffy (von Bertalanffy, 1972) added interrelations within the different systems.

Open systems are free of boundaries and have no controlling mechanisms, they operate freely and are self-organizing. Open systems exchange information, resources, and energy freely (Turner and Baker, 2019; Kast and Rosenzweig, 1972). In contrast, closed systems are bounded systems with a clear separation from external/environmental forces. The boundaries placed around a closed system help to protect the system and its components (systems, sub-systems) from external/environmental forces. There are also some systems that could be classified as being both closed and open (Kast and Rosenzweig, 1972), this depends on the kind of boundaries placed on the system, or with a system that is partially controlled with other unprotected systems (open).

Systems theory has been applied to some social systems that are considered both open and closed systems. Others make the claim that systems theory can be applied to purely open, social systems. This was cautioned against by a number of researchers, however, because systems theory requires a set of boundaries to function well. The problem with some social systems is that they often have no clear boundaries. Without clear boundaries, it is hard to identify the system or any of its component systems. Applying boundaries to social systems can be difficult, as with an organization (Wang, 2004). Another disadvantage of applying systems theory to social systems is that the process becomes too mechanistic: "Systems theory is indeed the ultimate step toward the mechanization and devaluation of man and toward technocratic society" (von Bertalanffy, 1972: 423–424). Additional problems occur when you add the human element where humans have free will (Kast and Rosenzweig, 1972) and do not operate within a set of predefined boundaries. Social systems have purpose and have the potential to evolve beyond the boundaries of the systems or system, adding a level of unpredictability (Kast and Rosenzweig, 1972; Turner and Baker, 2019).

All systems have common laws among them, which is what general system theory is all about, identifying general laws that apply to all, or most, systems. In these closed systems, traditional cause-and-effect relationships exist, and root cause analysis can be determined using problem-solving models like the plan-do-check-act cycle.

System relates to the whole. This whole is composed of a number of different systems, each with their own set of subsystems. Interconnections exist between the subsystems, between some subsystems and the systems, and between some of the systems with the system. Feedback loops are put in place as a means of controlling the subsystem or the systems. This control function, however, only works for closed systems. When operating open systems, it is difficult to control the system, or the functions of the system, through traditional control mechanisms. Operating open and social systems requires a different type of framework and a change in thinking.

Complexity Theory

In closed systems, interactions between subsystems are controlled as a means of maintaining the system. In open systems, the interactions occur freely. These interactions are not controlled but rather come from self-organizing functions that are learning and adapting to meet environmental demands. Open systems learn, adapt, and transform themselves to new or modified entities designed to meet the challenges imposed on them by their environment. These transformations are identified as emerging from chaos or complexity, and this process is termed emergence. The interactions of these open systems that lead to emergence can be viewed as the antecedents. Emergence results in new outcomes not previously predicted or expected. It is these interactions, internally and externally, that drives a system to coevolve and emerge into new states.

Complexity theory comes from the field of complexity science that is best described by the following description:

> Complexity science targets a sub-set of all systems; a sub-set which is abundant and is the basis of all novelty; a sub-set which is evidenced in biology, chemistry, physics, social, technical and economic domains; a sub-set which coevolves with its environment; a sub-set from which structure emerges. That is, self-organization occurs through the dynamics, interactions and feedbacks of heterogeneous components. . . . This sub-set of all systems is known as complex systems. (Strathern and McGlade, 2014: 12)

With complexity theory, simple systems can result in complex behaviors, while, at the same time, complex systems can result in simple behaviors (Gleick, 2008). An example of the former is when a small virus goes undetected and rapidly turns into an epidemic, or even worst, a pandemic. The severe acute respiratory syndrome (SARS) epidemic that took place in 2003 is one recent example of an epidemic, whereas the HIV/AIDS virus is an example of a recent pandemic (WebMD, 2019). In contrast, the latter can be found in our cognitive functions. The human brain is definitely a complex system; however, it produces simple outcomes, such as reading, writing, communicating, and solving problems, to name only a few.

The laws of complexity hold true, just as common laws of systems hold true for GST. The laws of complexity hold true universally, however (Gleick, 2008). The system holism rule, that the whole is greater than the sum of its parts, also changes in that the whole often evolves into a new entity or unit because of emergence. From this, the system holism rule for complexity changes to the whole is "qualitatively different from their parts. . . . They cannot be meaningfully compared-they are different!" (Richardson, 2004: System Holism; Turner and Baker, 2019). A simple example of this can be found in individual learning. As a person grows, gains knowledge, and gains experience, learning occurs. Through this learning process, which some may call maturing whereas others may call it pursuing wisdom, the individual changes. From a systems perspective, the individual is a composite of the knowledge and experiences endured. This is not exactly correct in that the person, the whole, is also different. This would be the perspective taken from complexity: the whole (individual) is qualitatively different, in part because the parts also have changed. It is a dynamic process that changes over time. In contrast, systems theory views the whole mostly as a constant, or at least it is preferred to remain unchanged.

Complexity theory can be defined in a number of different ways. One definition relates to studying patterns: "A study of changing patterns or order, self-organization, or constrained diversity" (Anderson et al., 2012: 964). Other definitions relate to agents involved in complex environments: "Studies the behavior of complexity interacting, interdependent, and adaptive agents under internal and external pressures" (Borzillo and Kaminska-Labbe, 2011: 355). Other definitions identify the elements of complexity: "Made up of a very large number of autonomous elements. . . . Dynamic, interactive, governed by micro-rules, exhibit 'butterfly effects', non-linear, and exhibit replicated patterns" (Campbell-Hunt, 2007: 796–797).

Complexity theory can be divided into two components, chaos theory and complex adaptive systems (CAS; see Figure 4.1). Chaos theory generally identifies order from disorder as a means of understanding complexity. Chaos theory encompasses multiple disciplines (e.g., physics, biology, mathematics, social science) and determine new laws, models, and formulae to describe disorder. This field of science is truly a transdisciplinary field of science, requiring new tools and methods to be created to explain disorder. Within chaos theory, scientists believe "that simple, deterministic systems could breed complexity; that systems too complex for traditional mathematics could obey simple laws; and that, whatever their particular field, their task was to understand complexity itself" (Gleick, 2008: 307). What is portrayed in this quote from Gleick is that, through mathematics, the field of chaos tries to develop new formulas that explain the chaotic phenomenon. It is believed, in general, that through mathematic formulas, chaos can be explained.

CAS are self-organizing open systems, which sets them apart from chaotic systems, and generally are described in the following definition: "Open dynamical systems that are able to self-organize their structural configuration through

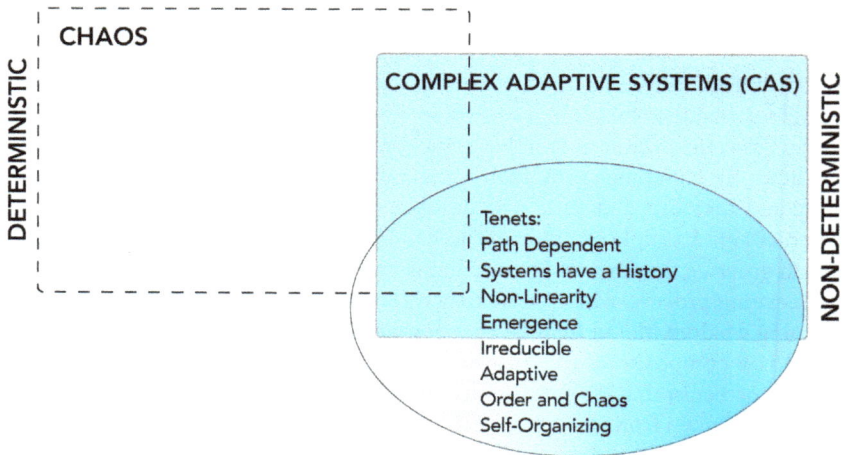

FIGURE 4.1. Complexity Theory

the exchange of information, energy and other resources within their environment, are able to transform these resources in order to support action" (Larson, 2016: Firms). In this definition, new entrepreneurial startups would be a CAS. New startup companies utilize information, energy, and resources (e.g., capital from external sources) to deliver value to the customer. New startup companies self-organize as they expand and grow into new markets. These self-organizing functions provide the new startup with the structure that it needs to adapt and operate in new markets.

CASs have cycles of being stable (order) and unstable (disorder) as they react to environmental forces. These cycles represent the systems' learning processes, suggesting that CASs are capable of learning to adapt to the environment to achieve a sense of order, if only temporary. This separates CAS from chaotic systems; however, both are used to explain complexity. These differences highlight the distinction between levels of complexity that are present within a particular system at any point in time. Pure complex systems could be viewed as being a chaotic system. The two have some overlapping characteristics (at the high end of complexity and low end of chaos; see Figure 4.1). As one moves away from chaotic systems, however, complexity can be explained from the perspective of CAS. In this regard, CASs are viewed as having cycles of order and disorder, compared with its counterpart, chaos, which is always in disorder.

The discussion of CAS continues later in this chapter and will be the focus of TFS. CASs have been associated with a number of social systems, such as colleges and universities (Davis et al., 2015), communities of practice (Borzillo and Kaminska-Labbe, 2011), emergency responders and trauma centers (Beck and Plowman, 2014; deMattos et al., 2012), entrepreneurship (Anderson et al.,

2012; Crawford and Kreiser, 2015), organizational learning (Antonacopoulou and Chiva, 2007; Chiva et al., 2014), project management (Aritua et al., 2009), strategic management (Bovaird, 2008; Burgelman and Grove, 2007; Campbell-Hunt, 2007; Tong and Arvey, 2015), supply chain management, and risk management (Hearnshaw and Wilson, 2013; Manuj and Sahin, 2011). Organizations also have been associated with being a CAS (Aagaard, 2012; Boal and Schultz, 2007).

In its most basic form, we have learned the following basic laws of complexity: "Simple systems give rise to complex behavior. Complex systems give rise to simple systems. And most important, the laws of complexity hold universally, caring not at all for the details of a system's constituent atoms" (Bode and Wagner, 2015: 304).

These basic laws of complexity set complexity theory (chaos and CASs) apart from general system theory and systems theory. Figure 4.2 shows the differentiation between complexity theory and system-theoretical approaches. Figure 4.2 also illustrates the overlap present between CAS and system-theoretical approaches. This overlap is due, in part, to system-theoretical approaches being applied to open systems. If the level of complexity is low, some of these system-theoretical approaches may work. When complexity is high, however, these techniques are ill-equipped to address nonlinearity, uncertainty, and irreversible processes. When medium to high levels of complexity exist, it is beneficial to view the whole as a CAS, beginning the transition toward complexity thinking.

Complexity thinking is presented in a later chapter. Currently, it is important to understand that system-theoretical approaches are successful for specific types of problems and complexity thinking approaches are successful for different types of problems. Also, neither are truly required or applicable to all types of problems. For example, system-theoretical approaches have been shown to be quite effective in managing manufacturing facilities. This process is more linear and would not need to be managed or evaluated using complexity theory that

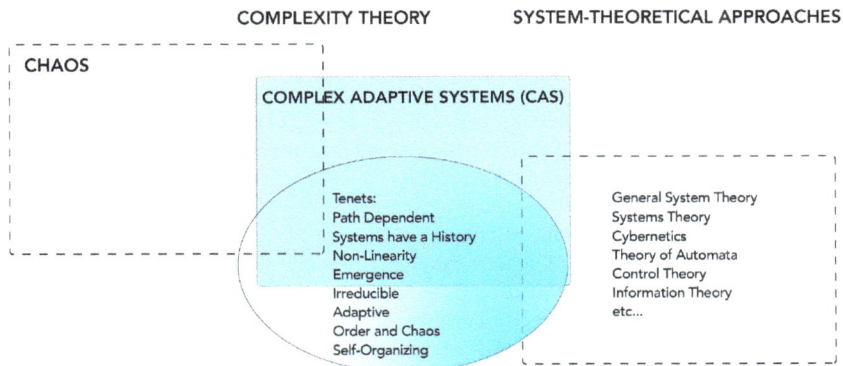

FIGURE 4.2.   Complexity Theory and System-Theoretical Approaches

looks at nonlinear system. In contrast, many economic theories have been unreliable in that they fail to produce reliable predictions. Many economic theories are based on what is called perfect rationality that assumes "agents have the virtue of being perfectly predictable . . . they know everything . . . and they use flawless reasoning to foresee . . . their actions" (Waldrop, 1992: 141). In applying concepts of complexity, and knowing that humans are not always perfectly rational, and that the economy is a CAS, economic theories are beginning to be challenged. "In nonlinear systems—and the economy is most certainly nonlinear—chaos theory tells you that the slightest uncertainty in your knowledge of the initial conditions will often grow inexorably. After a while, your predictions are nonsense" (Waldrop, 1992: 143).

It is important to know what type of problem is being addressed before deciding which method or tool to select. This comes with knowing more about complexity and complex problems. The following chapter, Chapter 5, Complexity, talks more about complex problems and introduces different tools and techniques to the reader for navigating the waters of complexity. Chapter 6, Systems Thinking Versus Complexity Thinking, contrasts systems thinking with complexity thinking and introduces the characteristics of CAS.

References

Aagaard P. (2012) The challenge of adaptive capability in public organizations: A case study of complexity in crime prevention. *Public Management Review* 14: 731–746.

Anderson AR, Dodd SD, and Jack SL. (2012) Entrepreneurship as connecting: some implications for theorising and practice. *Management Decision* 50: 958–971.

Antonacopoulou E and Chiva R. (2007) The social complexity of organizational learning: The dynamics of learning and organizing. *Management Learning* 38: 277–295.

Aritua B, Smith NJ, and Bower D. (2009) Construction client multi-projects: A complex adaptive systems perspective. *International Journal of Project Management* 27: 72–79.

Beck TE and Plowman DA. (2014) Temporary, emergent interorganizational collaboration in unexpected circumstances: A study of the *Columbia* space shuttle response effort. *Organization Science* 25: 1234–1252.

Boal KB and Schultz PL. (2007) Storytelling, time, and evolution: The role of strategic leadership in complex adaptive systems. *The Leadership Quarterly* 18: 411–428.

Bode C and Wagner SM. (2015) Structural drivers of upstream supply chain complexity and the frequency of supply chain disruptions. *Journal of Operations Management* 36: 215–228.

Borzillo S and Kaminska-Labbe R. (2011) Unravelling the dynamics of knowledge creation in communities of practice though complexity theory lenses. *Knowledge Management Research and Practice* 9: 353–366.

Bovaird T. (2008) Emergent strategic management and planning mechanisms in complex adaptive systems: The case of the UK best value initiative. *Public Management Review* 10: 319–340.

Burgelman RA and Grove AS. (2007) Let chaos reign, then rein in chaos-repeatedly: Managing strategic dynamics for corporate longevity. *Strategic Management Journal* 28: 965–979.

Campbell-Hunt C. (2007) Complexity in practice. *Human Relations* 60: 793–823.

Chiva R, Ghauri P, and Alegre J. (2014) Organizational learning, innovation and internationalization: A complex system model. *British Journal of Management* 25: 687–705.

Cognitive Edge. (2010, August 24) *Summary article on origins of Cynefin.* Available at: https://cognitive-edge.com/articles/summary-article-on-cynefin-origins/.

Crawford GC and Kreiser PM. (2015) Corporate entrepreneurship strategy: extending the integrative framework through the lens of complexity science. *Small Business Economics* 45: 403–423.

Davis AP, Dent EB, and Wharff DM. (2015) A conceptual model of systems thinking leadership in community colleges. *Systemic Practice and Action Research* 28: 333–353.

deMattos PC, Miller DM, and Park EH. (2012) Decision making in trauma centers from the standpoint of complex adaptive systems. *Management Decision* 50: 1549–1569.

Gleick J. (2008) *Chaos: Making a new science.* New York, NY: Penguin Books.

Hearnshaw EJS and Wilson MMJ. (2013) A complex network approach to supply chain network theory. *International Journal of Operations and Production Management* 33: 442–469.

Kast FE and Rosenzweig JE. (1972) General system theory: Applications for organization and management. *Academy of Management Journal* 15: 447–465.

Kurtz CF and Snowden DJ. (2003) The new dynamics of strategy: Sense-making in a complex and complicated world. *IBM Systems Journal* 42: 462–483.

Larson CS. (2016) Evidence of shared aspects of complexity scinece and quantum phenomena. *Cosmos and History: Journal of Natural and Social Philosophy* 12: 160–171.

Manuj I and Sahin F. (2011) A model of supply chain and supply chain decision-making complexity. *International Journal of Physical Distribution and Logistics Management* 41: 511–549.

Richardson K. (2004) Systems theory and complexity: Part 1 [Forum]. *Emergence: Complexity and Organization* 6: 75–79.

Roberts N. (2000) Wicked problems and network approaches to resolution. *International Public Management Review* 1: 1–19.

Schon DA. (1983) *The reflective practitioner: How professionals think in action.* New York, NY: Basic Books.

Strathern M and McGlade J. (2014) *The social face of complexity science: A festschrift for professor Peter M. Allen,* Litchfield, AZ: Emergent Publications.

Tong YK and Arvey RD. (2015) Managing complexity via the Competing Values Framework. *Journal of Management Development* 34: 653–673.

Turner JR and Baker R. (2019) Complexity theory: An overview with potential applications for the social sciences. *Systems* 7: 23.

von Bertalanffy L. (1968) *General system theory: Foundations, development, applications.* New York, NY: George Brazillier.

von Bertalanffy L. (1972) The history and status of general systems theory. *Academy of Management Journal* 15: 407–426.

Waldrop MM. (1992) *Complexity: The emerging science at the edge of order and chaos.* New York, NY: Simon & Schuster Paperbacks.

Wang T-W. (2004) From general system theory to total quality management. *Journal of American Academy of Business* 4: 394–400.

WebMD. (2019) *Pandemics: Epidemics, pandemics, and outbreaks.* Available at: https://www.webmd.com/cold-and-flu/what-are-epidemics-pandemics-outbreaks#1.

Yawson RM. (2012) Systems theory and thinking as a foundational theory in human resource development: A myth or reality? *Human Resource Development Review* 12: 53–85.

CHAPTER 5

Complexity

Complexity occurs in agents, organisms, and systems in different ways, at different times, and at various levels of analysis. For example, look at the rapid growth of autism in the United States. One could tell that autism occurs in some people (agents) and not others and that autism varies as measured by the autistic spectrum disorder (ASD). This trend has increased over the years, the Center for Disease Control and Prevention (CDC) estimated that, in 2018, 1 in 59 children were diagnosed with an ASD. The causes vary as do the symptoms, but we have been fortunate enough to have an early diagnosis of ASD in children before the age of three. ASD is clearly a complex problem that has surfaced, not only in the United States, but globally.

This incongruence of not knowing, not understanding, and not being able to predict, results in a constant feeling of uncertainty with the feeling that it is nearly impossible to deal with complexity. Fortunately, the field of complexity has been studied for some time, and we are now capable of providing some guidance. One benefit is in identifying different types of complex problems, in part by identifying the level of complexity that is present in the environment. Another benefit comes in the steps that can aid one when addressing these complex problems. The following sections provide information about identifying the different types of complex problems followed by steps to address these problems.

Complex Problems

Complex problems originally was identified by Churchman (1967) when discussing simple, complex, and wicked problems. This distinction of problem classification had carried over to three distinct types of problems:

- Type 1 problems represent simple problems and are defined as having consensus on the problem and solution.
- Type 2 problems are complex problems that have an agreed upon problem but no agreed upon technique to solving the problem.
- Type 3 problems represent wicked problems in which neither the problem nor the solution can reach any consensus on what the problem is or how it can be solved. (Roberts, 2000)

Xiang (2013) highlighted the following five characteristics in which wicked problems could be identified. These five characteristics are summarized as follows:

- *Indeterminacy in problem formulation*—the precise formulation of a wicked problem as a problem with unique and determinate satisfaction conditions is virtually impossible because the values and interests of concerned and affected parties are diverse, often in conflict with one another, and change over time and across generations.
- *Nondefinitiveness in problem solution*—a rigorous and ultimate solution to a wicked problem with definitive results is unattainable because neither the problem nor the repercussions of its solution are determinate.
- *Nonsolubility*—wicked problems can never be solved because of the first two characteristics. Unlike "tame problems" that are determinate with clear goals(s) and a definite set of well-defined rules (like those in mathematics, engineering, and chess), and are thus ultimately soluble (eliminable), wicked problems may be suppressed or even overcome, but cannot be eliminated. In different and often more wicked forms, they will recur.
- *Irreversible consequentiality*—every implemented solution to a wicked problem is consequential, often triggering ripple effects throughout the entire socio-ecological system that are neither reversible, nor stoppable.
- *Individual uniqueness*—despite likely similarities among wicked problems, there always is one or more distinguishing property of overriding importance that makes an individual problem and its solution(s) essentially *one-of-a-kind*. There are therefore no classes of wicked problems, nor immediately transferable solutions. (Xiang, 2013: 1)

The distinction between problems is an important one in that it will dictate how the problem can be addressed. The techniques, methods, manpower, and available resources will partially depend on the types of problems that are being experienced. Although Roberts (2000), Xiang (2013), and others classify wicked problems as the highest level of complexity and uncertainty, we conceptualize problems based on the level of complexity perceived. We revert to the level of complexity, with low to moderate levels of complexity being equivalent to Roberts (2000) Type 2 problems, and high levels of complexity being equivalent to Roberts (2000) Type 3 problems.[1] Identifying the type of problems that typically

---

1   Type 1 problems are simple problems that have no characteristics of complexity and traditional methods of solving them are adequate. We are interested mostly in Type 2 and 3 problems that involve complexity.

are experienced in any team or organization will help determine the type of team or organizational structure that will be required to address these problems. Later, these problems will be matched with the Cynefin framework (Kurtz and Snowden, 2003) to identify not only the type of problem that is being experienced but also the different techniques that can be utilized to help solve and resolve these problems.

Addressing Complex Problems

When viewing simple or complicated problems, it is easy to break down the problem into simple microproblems. By attacking each microproblem, one can combine the results to get to the initial problem. These practices typically work well with simple or complicated problems because the problem, and the variables, can be defined and measured. In addition, different techniques for solving the problem also can be identified.

For simple and complicated problems, these reductionistic techniques, breaking problems into subproblems, have been with us for some time and have become one of the more common practices used in research. Reductionistic methods identify correlations between variables and highlight which variable could be affecting other variables. Hypothesis testing is conducted and symmetric relationships are identified from the analysis. These symmetric relationships are often the result of statistical tests (e.g., analysis of variance, ANOVA; multiple regression and correlation, MRC; structural equation modeling, SEM) in which one or more independent variables (IV; X1, X2, X3) are found to be associated (either positively or negatively) with a dependent variable (DV; Y).

Unfortunately, results from such hypotheses testing have been identified as having multiple problems, such as not being normally distributed (symmetrical), having an inadequate sample size, being a cross-sectional design (only one point in time measurement), or making the results invalid or irrelevant because the tested theoretical model is ill-defined. In addition, "offering hypotheses and testing for symmetric relationships alone is usually insufficient for understanding, describing, and predicting relationships between simple/configural X and simple/configural Y relationships" (Woodside, 2017: vi). A better explanation, understanding, or sensemaking is needed here: "What is needed in science is not just facts but relevant facts" (Chalmers, 2013: 25).

These reductionistic practices have been shown to provide only a paucity[2] of understanding. Although providing some knowledge and understanding of the world around us, they do, however, have their limitations. The main issue, when viewing complex problems, is that complexity does not occur in a vacuum, and boundaries cannot be placed around the problem to test the hypothesis in real

---

2   Paucity, in this context, refers to having a small amount of understanding (small, limited, scarce).

time. Other issues include the fact that complexity is nonpredictable and unexpected outcomes can occur.

Complexity Science

To address these issues of complexity, a new science has been formed called complexity science. Complexity science expands on the reductionistic practices by looking not only at the problem and its subproblems but also at the interactions between each subproblem as well as how these interactions affect or change the behavior of the whole. Complexity science results in "having a more comprehensive and complete understanding of the whole" (Turner and Baker, 2019: 2). Complexity science also looks at unbounded and open systems as opposed to bounded and closed systems.

Having open systems provides a system in which all relevant influential variables can be observed at the same time. Looking for patterns, or lack of patterns, is the object when viewing these open systems. This has led to new techniques in research to identify these patterns. Some examples include network analysis techniques, computer automata, automated intelligence, and machine learning.

An example that shows these two different perspectives in action, reductionism and connectives, can be found in Roger Lewin's book *Complexity: Life at the Edge of Chaos*. Lewin described an experience that Jim Lovelock[3] had when consulting for NASA in their quest to determine whether there was life on Mars:

> Jim was hired by the National Aeronautical and Space Administration (NASA) as a consultant in their quest to discover whether there was life on Mars. NASA's idea was to look directly for signs of life on the planet's surface: microscopically, looking for microbe-shaped objects; and chemically, seeking signs of microbial metabolism of the sort biologists are familiar with on Earth. Jim considered this a chancy approach, and hit upon a more global view. If the planet were dead, he reasoned, then its atmosphere would be determined by physics and chemistry alone; it would be in equilibrium with the chemistry of the planet's minerals. But if life lurked there, however simple, it would undoubtedly exploit the atmosphere for raw materials, thus changing its chemical composition. A living planet would have an atmosphere shifted away from a simple equilibrium with chemistry and physics of rocks. Simple argument; compelling strategy; ignored. NASA chose the chemical analysis route, and when the Viking lander sent back results in 1975, they were ambiguous at best. (Lewin, 1992: 112)

This example shows both perspectives in action. The reductionistic chemical process chosen by NASA produced ambiguous results. This partially was due to using the wrong method because of the level of complexity involved. The method

---

3   Jim Lovelock is the creator of the Gaia theory, which states the following: "All living and non-living components on earth work together to promote life" (Ravilious, 2008).

proposed by Jim Lovelock would have taken complexity into account by looking for and identifying patterns. The connectionist method was conducted by Jim Lovelock with satisfactory results. Jim was quoted in Lewin's book as saying: "There's no life on Mars. . . . I knew that from the spectral analysis of the atmosphere" (Lewin, 1992: 112).

Organizational Transformation

It wasn't too long ago when management theory and practice was called upon to shift in order to plan for the future. Deming's original call stated: "Transformation of American style of management is not a job of reconstruction, nor is it revision. It requires a whole new structure, from foundation upward" (2000: ix). At the time, this call was made to American management and government agencies by Deming in his book *Out of the Crisis* (first published in 1982). Deming indicated that the basic problem, in part, was due to top managements' failure to manage (Deming, 2000). Today, Deming's call could be replicated to push for this new managerial flow to meet the needs of globalization, complexity, software, and artificial intelligence, to name but a few. Today's call, however, extends beyond just American management and organizations; this call is for all globalized partners, including for-profit and not-for-profit institutions. This transformation is a globalized transformation and must be achieved through both bottom-up and top-down processes working toward similar goals.

The concept of organizational transformation has shifted from one that previously focused on mass production and bureaucracy through "routinization, standardization, control, and automation" (Orlikowski, 1996: 63) to one that involves complexity in which "visions of agile manufacturing, virtual corporations, and self-organizing teams are prominent" (Orlikowski, 1996: 63). In this new era, organizational transformation is achieved through constant change, whereas stability and operating as usual is out (Orlikowski, 1996).

Why is change so hard? It is hard because it involves people changing their behaviors, and in some cases, relinquishing control or power. Change occurs through a number of different mechanisms; the two prominent types are planned and emergent change.

Planned Versus Emergent Change

Planned change is described best as change directed by managers that are designed to meet organizational performance measures. In contrast, emerging change occurs when one realizes new patterns of organizing (Orlikowski, 1996) absent of predefined intentions or biases. Emergent change is reflective of the decisions made by the self-organizing frontline employees. Managements' responsibility is to identify the patterns from these decisions and to make sense of what these patterns mean:

The job of management is to author interpretations and labels that capture the patterns in those adaptive choices. Within the framework of sensemaking, management sees what the front line says and tells the world what it means. In a newer code, management doesn't create change. It certifies change. (Weick, 2009: 239)

Because emergent change can only be realized (from identification of patterns), sensemaking is essential (Weick, 2009). Sensemaking is achieved through the following four basic conditions: "(1) stay in motion, (2) have a direction, (3) look closely and update often, and (4) converse candidly" (Weick, 2009: 235).

Although planned change may work in simple systems, it rarely works in complicated or complex systems. Change is a definite given enough time. Change is also more prevalent and occurs more frequently in open systems than in closed systems. Control or filter mechanisms are introduced to closed systems to protect the system from external or environmental disturbances, whereas open systems are free of these control or filter mechanisms, making them capable of adapting to external or environmental perturbations.

When addressing change in closed systems, one can typically manage using planned change while adapting to variances within the system to maintain equilibrium. In contrast, when functioning change in open systems, planned change is not enough to survive. Plans are disrupted because of the constant bombardment of external and environment forces as well as from changes within. Open systems must be capable of recognizing and managing emergent change, this includes constantly updating strategy and plans. Here, change is viewed as being "ongoing, continuous, and cumulative" (Weick, 2009: 230).

The concepts of planned and emergent change will be discussed again after the introduction of the Cynefin framework.[4] But first we will expand on the concept of sensemaking.

Sensemaking

In most cases, it is easy to make sense of the parameters surrounding simple and complicated problems. When the parameters are unclear, it is fairly easy to involve an expert to describe the parameters and to situate the problem for the setting (to contextualize the problem). Making sense of simple and complicated problems is possible because of the understanding of the system in which the problem resides. For example, in manufacturing, when a problem occurs, it can be narrowed down to the process or subprocess responsible for the phase in the product's development in which the defect was found. Through cause-and-effect methods, the root cause of the problem can easily be identified.

---

4   Cynefin (pronounced kuh-nev-in) is a Welsh term translated to mean habitat or place. Cynefin relates to a place in which we can never be fully aware of our surroundings, and that our surroundings definitely have an influence on us (Kurtz and Snowden, 2003).

This trouble-shooting process is successful because of the knowledge of the system, or the manufacturing process, as in this simple example.

When dealing with complex problems or environments, the parameters are unknown and do not make sense, even to most experts. A good example is in most election processes. It is impossible to know which candidate will win or who has the advantage at any one time. In hindsight, these parameters may become obvious, but they are unclear as the events unfold in real time.

Sensemaking is a tool used to help make sense of such complexity. Sensemaking, as a tool, is composed of collecting stories from a sample (agents) and interpreting these stories to identify specific patterns that may lead to better understanding of the complexity involved. Stories and narratives are essential elements to the sensemaking process.

Karl Weick coined the term sensemaking, or at least brought it into view. Weick described sensemaking as follows: "The interplay of action and interpretation rather than the influence of evaluation on choice" (Weick, 2009: 132). Sensemaking requires narratives, or storytelling, for people (agents) to conceptualize their surroundings and to form their own set of mental models. In the following excerpt, many of the characteristics identified in the "Complexity Thinking" section also are touched upon (e.g., flow, unknowable, stories, and emergence).

> To focus on sensemaking is to portray organizing as the experience of being thrown into an ongoing, unknowable, unpredictable streaming of experience in search of answers to the question, "what's the story?" Plausible stories animate and gain their validity from subsequent activity. The language of sensemaking captures the realities from agency, *flow*, equivocality, transience, reaccomplishment, unfolding, and emergence, realities that are often obscured by the language of variables, nouns, quantities, and structures. (Weick, 2009: 132; empahsis added)

Sensemaking is an extension of one's knowledge, experiences, education, culture, and the contextual setting. What this means is that, regardless of one's capabilities, a person who is capable of making sense of complex and difficult issues or situations is more likely to be successful in resolving these issues than someone who is unable, or unwilling, to make sense of these issues.

As humans, we tend to favor ordered and simple rules. This organization of our environment leads us to ordering things into simple and understandable rules. Unfortunately, as things become more complicated or complex, our ability to obtain this sense of order becomes diminished. Having the ability to practice sensemaking techniques helps us make sense of these chaotic events.

Describing what sensemaking is, Weick provided the following:

- Sensemaking organizes flux.
- Sensemaking starts with noticing and bracketing [mental models].
- Sensemaking is about labeling.
- Sensemaking is retrospective.

- Sensemaking is about presumption.
- Sensemaking is social and systemic.
- Sensemaking is about action.
- Sensemaking is about organizing through communication. (Weick, 2009: 134–137)

Sensemaking begins with chaos, or a chaotic event (increasing level of complicated or complex events). Without a chaotic event, sensemaking is not necessary, everything is already in order. As an environment becomes more chaotic, it is essential to notice abnormal signals (weak signals) as they occur and to develop new meaning surrounding these weak signals. Weick (2009) termed this practice noticing and bracketing; today, we use the term developing mental models or shared cognitive models.

Labeling is a form of organizing, or way of knowing, that has become common practice. This practice induces meaning that does not extend beyond ourselves (the human race): "Social objects and phenomena such as 'the organization', 'the economy', the market' or even 'stakeholders' or 'the weather', do not have a straightforward and unproblematic existence independent of our discursively-shaped understandings" (Chia, 2000: 513). Labels are constructed and become fixed to communicate them socially (Chia, 2000). This process of constructing labels or meaning, a process of sensemaking, is described in the following: "It is through this process of differentiating, fixing, naming, labelling, classifying and relating-all intrinsic processes of discursive organization-that social reality is systematically constructed" (Chia, 2000: 513).

To have comprehensive knowledge, one must understand the history that led to the chaotic events or weak signals. Reflection is required to look back on the sequence of events to determine what changes were made and when. Also, when an action is taken, retrospectives are essential. Retrospectives are after-action reviews. Retrospectives act not only to review what had been done but also to discuss the next course of action and to identify whether or not any errors had occurred during the action. Reflection, and retrospectives, are part of the learning process and are critical components of sensemaking.

At some point, the process of sensemaking needs to transition from the level of abstraction to being concrete and contextual (situational and specific to location and agents). Here, the mental models that highlighted the weak signals are transferred to actual activities or events that potentially caused the weak signals to surface. Once concrete examples or events have been identified, sensemaking transitions into action to tame the chaos. This action, along with communicating the mental models and level of concreteness, is a social activity that is another step in the sensemaking process. These processes are systemic in that changes or new policies made to tame the chaos may have additional influence on other agents or processes. The systemic impact of any change needs to be considered throughout the sensemaking action phases and also must be communicated to those involved.

Sensemaking does not necessarily require all of the steps to be practiced. Sensemaking, however, cannot be achieved by implementing one single step, as most of the steps are required to achieve understanding. For example, constructing a label or category for an object or event alone does not constitute meaning— this is where other components of sensemaking are essential in constructing meaning. Sensemaking acts as a technique to "translate the difficult and the intransigent, the remote or resistant, the intractable or obdurate into a form that is more amenable to functional deployment" (Chia, 2000: 517). More important, sensemaking aids in understanding complexity by identifying weak signals, organizing these signals into mental models and labels, communicating these mental models as a means of social communication, and providing opportunities to develop theories and to test these mental models as a means of acting on these weak signals.

Another aid in achieving sensemaking capabilities is the utilization of the Cynefin framework, which aids in the development of sensemaking in organizations (Cognitive Edge, 2010).

The Cynefin Framework

Random Boolean Networks

Random Boolean networks, from the science of biology and evolutionary science, represent vast families of disordered networks operating in one of two possible states: active or unfrozen, and inactive or frozen. In addition to being either active or inactive, these networks operate in one of three types of behavior: ordered, complex, and chaotic (Kauffman, 1993). These networks function similar to the following summary:

> In the *ordered* regime, many elements in the system freeze in fixed states of activity. These frozen elements form a large connected cluster, or *frozen component*, which spans, or *percolates*, across the system and leaves behind isolated islands of unfrozen elements whose activities fluctuate in complex ways. In the *chaotic* regime, there is no frozen component. Instead, a connected cluster of unfrozen elements, free to fluctuate in activities, percolates across the system, leaving behind isolated frozen islands. In this chaotic regime, small changes in initial conditions unleash avalanches of changes which propagate too many other unfrozen elements. These avalanches demonstrate that, in the chaotic regime, the dynamics are very sensitive to initial conditions. The transition from the ordered regime to the chaotic regime constitutes a phase transition, which occurs as a variety of parameters are changed. The transition region, on the edge of order and chaos, is the *complex* regime. Here the frozen component is just percolating and the unfrozen component just ceasing to percolate, hence breaking up into isolated islands. In this transition region, altering the activity of single unfrozen elements unleashes avalanches of change with a characteristic size distribution having many small and few large avalanches. (Kauffman, 1993: 174; emphasis added)

In this example, percolate refers to spreading throughout gradually, or to filter through. Here, having all frozen elements is best associated with order, having no frozen elements is best associated with chaotic, and having a mixture of both frozen and unfrozen elements is best associated with complexity. An example would be a nuclear power plant. The nuclear reaction inside the reactor's core is chaotic. The nuclear reaction involves no frozen elements because of the constant chemical reactions taking place inside the core to generate heat. The frozen elements include the control mechanisms that provide stability to maintain the nuclear reaction to remain at controllable levels. Order is maintained through the control mechanisms, frozen elements, while energy is produced from the heat generated by the nuclear reaction (unfrozen elements) taking place inside the core. A balance is maintained between order and chaos, preventing the nuclear reaction from getting out of control, which could result in total chaos. This is what happened with the 1986 Chernobyl accident and that nearly occurred earlier with the 1979 Three-Mile Island accident. In the latter, the reactor was controlled and in the former the reactor reached the state of disequilibrium and remains inoperable and radioactive to date.

The transitions just described, moving from order to complex, complex to chaotic, chaotic to complex, and complex to order, require an avalanche of change to take place before this phase transition can occur.

The Cynefin Framework

Previously, it was discussed that sensemaking is aided by the use of collecting narratives to develop mental maps, to create labels, and to communicate the abstract to more concrete levels of understanding. In talking about complexity, similar to the random Boolean networks just discussed, Kurtz and Snowden (2003) described a similar model within the context of organizational decision making and strategy formation called the Cynefin framework. The Cynefin framework identifies five distinct domains of knowledge: clear/simple/obvious,[5]

---

5   The initial Cynefin framework included the simple domain to represent known causes and effects. This simple domain represented best practices and instances in which one's cognitive load was capable of processing the tiny bit of information required to complete one's task (Kurtz and Snowden, 2003). Later, Dave Snowden changed this domain to the obvious domain representing control: "achieved through creating and mandating process, delegation of authority and so on" (Snowden, 2019: Planning and control), and again later to clear. We use clear for this domain (see Figure 5.1). Although some things may be obvious through mandated processes or through delegation of authority, other things may not be so obvious but remain simple to process. Take, for example, exiting a subway station. With the use of the exit signs (similar to job aids in the workplace), it is simple to find your way out of the subway. It may not be obvious, especially without the exit signs, but it is simple and clear. We are not taking a stand for either/or, but we will try to use clear in this book.

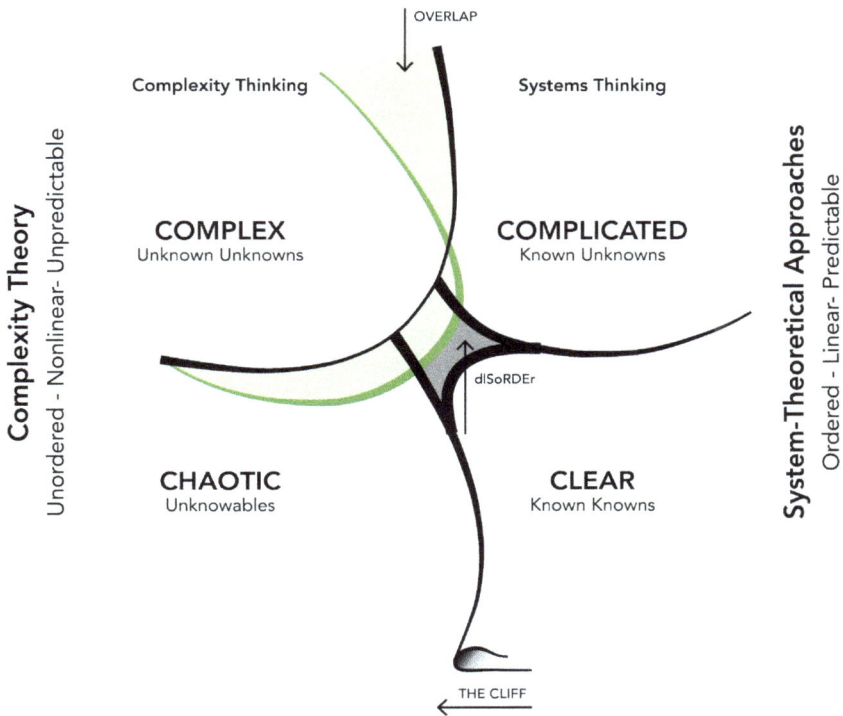

FIGURE 5.1.   The Cynefin Framework

Source: Cynefin™ *and the Cynefin™* framework are trademarks of Cognitive Edge Pte Ltd. and are used under license by Cognitive Edge.

known-knowns; complicated, known-unknowns; complex, unknown-unknowns; chaotic, unknowables; and disorder (Cognitive Edge, 2010; Kurtz and Snowden, 2003). Figure 5.1 provides an overview of the Cynefin framework. The Cynefin framework presents five different contextual settings to aid organizations and their leaders to "sense which context they are in so that they can not only make better decisions but also avoid the problems that arise when their preferred management style causes them to make mistakes" (Snowden and Boone, 2007: 70).

The first domain, originally known as simple and obvious, now called clear, represents consistently repeating patterns and events, clear cause-and-effect relationships, best practices,[6] and stability (Snowden and Boone, 2007).

---

6   The concept "best practices" is used frequently in the organizational literature and is discussed often in practice and in consulting. Note, however, that Toyota avoids the use of "best practices" because their culture is one of constant improvement. They believe that there is no best, only better. Best means we cannot improve, and this is never true. Best practices, as used in this book, represents the field and not Toyota's culture.

This represents the frozen components in the random Boolean network example. The simple domain is best represented by known-knowns and is characterized as being associated with the sense, categorize, and respond decision-making behaviors (Snowden and Boone, 2007; Snowden, 2012a).

The complicated domain requires more expert advice because of issues and problems that have multiple right answers and fuzzy cause-and-effect relationships (Snowden and Boone, 2007); it utilizes good, appropriate practices rather than best practices. The complicated domain is best represented by known-unknowns and is associated with the sense, analyze, and respond decision-making behaviors (Snowden and Boone, 2007; Snowden, 2012a).

The complex domain represents unpredictability and situations in flux in which there is no right answers and in which weak signals and emergent patterns need to be identified, requiring new creative and innovative techniques to function (Snowden and Boone, 2007). This domain is represented by unknown-unknowns and is associated with the probe, sense, respond decision-making behaviors (Snowden and Boone, 2007).

The chaotic domain represents high turbulence in which there is no clear cause-and-effect relationships with little to no time to make decisions (Snowden and Boone, 2007). This domain is represented as the unknowable domain (Snowden and Boone, 2007) and is associated with the act, sense, and respond decision-making behaviors (Snowden, 2012a).

In one sense, this model contrasts our level of certainty in our decisions with our level of certainty in our understanding of the situation (Kurtz and Snowden, 2003). The Cynefin framework began as a knowledge management model but later developed into much more. Not only had the Cynefin framework morphed into a significant framework for operating in various modes of knowledge (clear, complicated, complex, chaos, and disordered), but it also has been accepted as a tool for practicing sensemaking in the workplace (Snowden, 2012a). Through its maturation stages, the Cynefin framework has been influenced by, and has influenced, many essential institutions over the years (e.g., IBM, DARPA's Human Augmented Reasoning through Patterning, HARP).

The Cynefin framework is just that, a framework, as opposed to a categorization (Snowden, 2012b). Categorizations have predefined definitions and meanings with rigid boundaries. In contrast, a framework is less restrictive, is defined using narratives to construct its own meaning (sensemaking), and determines its own limitations and boundaries depending on the context. Frameworks also provide models that can be built upon to address different contextual situations. In contrast, when using categories, everything must fit neatly into the predefined buckets. Movement across each domain is easier to describe through sensemaking and self-constructivism techniques as opposed to using categorizations, as most examples would not fit into a predefined categorization model. Movement from one domain to the next is critical to identify, describe, and manage, as this movement is the key to creativity and innovation, to making sense out of complexity, and to managing in complex environments. The Cynefin framework provides such a system also known as Cynefin dynamics.

Movement from one domain to the next occurs naturally in society; however, being capable of managing these movements to more desirable outcomes is the key to operating in complexity. Understanding which moves are not warranted is also critical information to have. In general, movement goes from the simple domain to the complicated domain, from the complicated domain to the complex domain, and from the complex domain to the chaotic domain. Movement also can work in the opposite directions, from the chaotic domain to the complex domain, from the complex domain to the complicated domain, and from the complicated domain to the simple domain. Movement doesn't jump phases or domains. For example, one could not jump from simple straight to complex without first entering the complicated domain. The speed at which one travels across domains may vary, but skipping domains is not possible.

Although each domain is exclusive of the other domains, some level of overlap exists from one domain to the next. This overlap is best highlighted as *disorder* in Figure 5.1. This concept of disorder can be described as a level of fuzziness, not knowing the domain in which one is operating. Part of the sensemaking process is to collect enough data and information to determine the domain in which one is operating (i.e., to clear the fuzziness), and then domain-specific techniques can be utilized to address the problem. Movement, however, should not occur from the simple domain to the chaotic domain, nor from the chaotic domain to the simple domain. The description of a *cliff* is illustrated in Figure 5.1 to describe movement from the simple domain to the chaotic domain. The boundary between these two domains represents a cliff in which crossing this boundary can be quite costly.

Movement from one domain to the next does not occur without loss of energy. This transition is similar to the *avalanche of change* that is required to cross boundaries in the random Boolean networks previously described. Different domains have different combinations of frozen and unfrozen components. The energy expended to move from one domain to the next is high, and this energy must be transferred before being able to achieve the phase transition required to move. In addition, the methods and tools that can be utilized in one domain will not necessarily work in the other domains. For example, as best practices and job aids work well for the clear domain, these techniques will not work in the complex domain. A job aid that might work in the clear domain highlights specific procedures to follow. In contrast, the complex domain has too many unknowns, and an accurate job aid could not be developed. Even if a job aid could be developed, the conditions would change quickly, making the job aid irrelevant. Different methods and tools are required for each of the five domains.

Different Domains Require Different Tools and Methods

To highlight how different tools and methods are required for each of the five domains in the Cynefin framework, we start with an example of different improvement cycles for each domain. Change theory began with Lewin's (1951) three stages of change: unfreeze, change, and refreeze. Around the same time Deming introduced the plan-do-study-act (PDSA) cycle[7] for learning and improvement (Deming, 1994). This PDSA cycle later become popularized as the Deming cycle or the plan-do-check-act (PDCA) cycle (Liker, 2004). Deming's PDCA cycle had become the "cornerstone of continuous improvement" (Liker, 2004: 24) or *Kaizen* in Japanese and Lean terms. These cycles (such as PDCA) are capable of meeting the demands of frozen or ordered domains, and in some cases, can meet the demands of complex domains in which manageable levels of unfreezing can be controlled. These models, however, are not suited for complex domains with high levels of uncontrollable unfreezing islands, nor are they capable of addressing chaotic domains in which there is no freezing of any components.

Operating within the domains of high complexity and chaos requires a new mode of change. Here, inertia becomes problematic. Inertia identifies how an organization is unable to keep up with a changing environment, such as those entering into a complex environment. Operating in this domain, in the chaotic domain, often requires a revolution (Weick and Quinn, 1999), similar to what is known as the punctuated equilibrium theory,[8] which is defined as follows: "Radical and discontinuous change of all or most organizational activities is necessary to break the grip of strong inertia" (Romanelli and Tushman, 1994: 1143). These revolutionary changes also are linked to the "surrender of control" (Weick and Quinn, 1999: 373) from middle and upper management, leading to further resistance. Perhaps in the regions of chaos and in high levels of complexity, the trajectory of change should be considered being more "spiral or open-ended than linear" (Romanelli and Tushman, 1994: 382), or more in line with momentum.

---

7   Deming was influenced by the Shewhart Cycle (1939), which included specification, production, and inspection. Deming modified this model for the Japanese Union of Scientists and Engineers (JUSE) in 1950, a four-stage model consisting of design, produce, sell, and redesign. Further modification led to the Deming–Shewhart cycle that became known as the "Deming Wheel" in the Japanese market. The Deming Wheel included design, test in production, market, test in service, and redesign. From here, the Deming Wheel was coined. The PDCA cycle by the Japanese represents plan, do, check, and act. Although Deming stood firm on the PDSA cycle, the PDCA market became the model with which most people are familiar (Moen and Norman, 2009).

8   Punctuated equilibrium is a theory from biological evolution that claims organisms experience long periods of stability and lack of creativity until disrupted, or punctuated, by some change. This punctuated change is often disordered and unpredictable or chaotic (Godfrey-Smith, 2003).

The idea is to use momentum to move across domains or to move with a single domain. This momentum results in achieving flow—that is, flow that addresses change without disrupting value being delivered to the customer. Momentum also prevents the disruptive forces of inertia that wants to resist change completely.

One model of change that addresses complexity could be the observe-orient-decide-act loop, otherwise known as the OODA loop, initially the sense-orient-decide-act (SODA) loop. The OODA loop was developed by Col. John Boyd for the Air Force where he was tasked with determining the success experienced by F-86 pilots during the Korean War (Reinertsen, 2009). These pilots were able to "more quickly orient themselves to the tactical situation" (Reinertsen, 2009: 255) for a number of reasons. Aside from the advantages afforded by the aircraft design, the pilots were better able to act because of their abilities to make quick transitions, to change quickly, and to efficiently utilize their reserves (Reinertsen, 2009). These processes practiced in combat and emergency situations became known as the OODA loop (Figure 5.2).

The observe component of the OODA loop represents "assessing the environment, one's place in it, and the interaction of the two" (Boyd, 2018: 384). Orientation refers to a set of filters that influence one's sensemaking capabilities: culture, genetic heritage, new information, and previous experiences (Boyd, 2018). The orientation involves the analysis and synthesis stage, which results in an aggregate of all the orientation components in addition to any information obtained from the initial observation stage. The decision component results from the aggregate of the previous stages, including analysis and synthesis, resulting in a "process of projection, empathy, correlation, and selection" (Boyd, 2018: 385). The component of acting is a result of the decision stage, in which a hypothesis is presented to be tested—that is, a "hypothesis about how best to shape and to be shaped by the environment" (Boyd, 2018: 385). Identified as a nonlinear process, the OODA loop was described by Boyd as "an evolving, open-ended, far from equilibrium process of self-organization, emergence, and natural selection" (Boyd, 2018: 385). Figure 5.2 shows the OODA loop with each of the components just described. This figure also shows a number of feedback and feedforward mechanisms that provide an additional

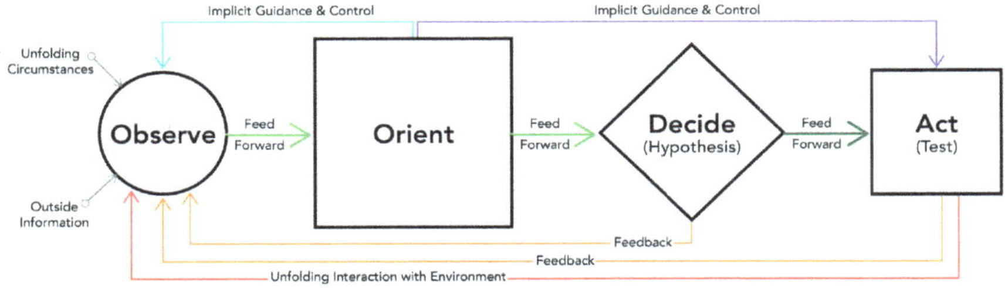

FIGURE 5.2. The OODA Loop

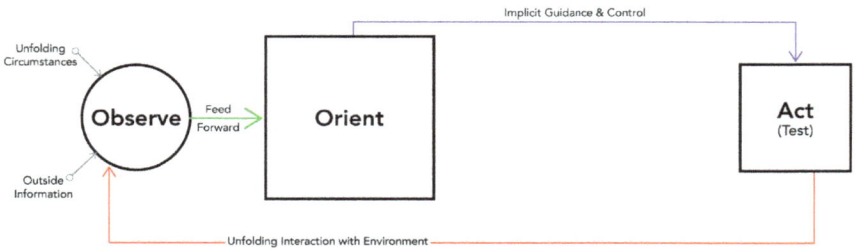

FIGURE 5.3.   Clear Loop

means for the process to be open and adaptive, as opposed to being closed and restrictive.

The OODA loop can work in various forms and has been connected to the Cynefin framework as one potential tool that operates across each domain. Earlier, each domain was represented as being composed of one form of decision-making behavior. For the clear domain, the process was characterized as being of the sense-categorize-respond type. The complicated domain was associated with the sense-analyze-respond type, the complex domain was associated with probe-sense-respond type, and the chaotic domain was associated with the act-sense-respond type. Each one of these four types of decision-making behaviors has an applicable version of the OODA loop.

Clear/Simple/Obvious Loop

The OODA loop relevant to the clear/simple/obvious domain is portrayed in Figure 5.3. This loop represents the sense, categorize, respond behavior (observe, orient, act in the OODA loop). In this domain, one begins with a quick observation of the environment, observe. In addition, what is captured cognitively during this period of observation is closely associated with one's filters (i.e., culture, genetic heritage, new information, and previous experiences). On the basis of what is observed and processed, the next step is action; no decision-making processes are required for the simple or obvious domain. This clear/simple/obvious loop is similar to single-loop learning, which occurs nearly automatically, in which an "error is detected and corrected without questioning or altering the underlying values of the system" (Tosey et al., 2011: 292).

Complicated Loop

The complicated loop is shown in Figure 5.4. This loop begins with one observing the environment and processing this information based on one's filters, much in the same manner as in the simple/obvious domain. Just as in the simple/obvious loop, the complicated loop does not necessarily require a decision stage,

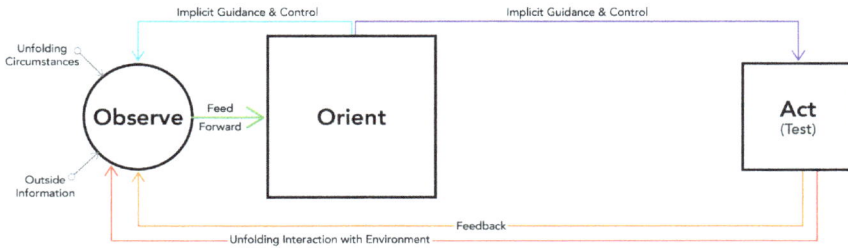

FIGURE 5.4. Complicated Loop

as this process is nearly automatic; known techniques and procedures can be followed to resolve complicated problems. This complicated loop is similar to double-loop learning in which errors are corrected by "first examining and altering the governing variables and then the actions" (Tosey et al., 2011: 292). Although single-loop learning does not require any analysis or synthesis, double-loop learning does have a stage in which analysis, synthesis, or both take place. This would be how the alterations are determined in the double-loop definition, alterations determined through analysis. This process also includes feedback loops to support the altering stages, further separating this complicated loop from the clear/simple/obvious loop.

Complex Loop

The next loop passes the threshold from the clear/complicated domains into the complex/chaotic domains. This threshold highlights the portion of complexity theory that is related to the left half of the Cynefin framework (unordered, non-linear, unpredictable; see Figure 5.1) with system-theoretical approaches (ordered, linear, predictable) on the opposite, right half of the Cynefin frame-work. The left half represents purely open and mostly open systems, whereas the right half represents nearly closed and fully closed systems. The complex loop extends from the complicated loop by incorporating the decision stage (decide; Figure 5.5). Not knowing all of the conditions, a decision needs to be made based on a quickly developed hypotheses, and this process can revert back to the observe stage (the first stage of feedback in Figure 5.5, decide to observe) to determine whether the initial hypothesis worked. This process is similar to developing prototypes when addressing complexity, given that prototypes provide hypotheses that can be tested in the environment and corrections and modifications can be made in response to what had been learned from the testing (the second stage of feedback in Figure 5.5; act to observe).

The complex loop is more of an exploratory loop because of the level of unknowns that are present. This loop is also similar to triple-loop learning. Triple-loop learning has been represented as coinventing or collective

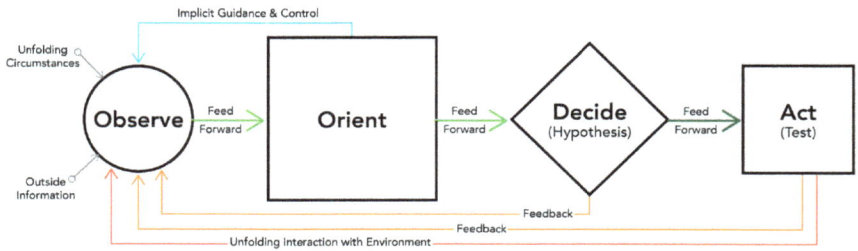

FIGURE 5.5.   Complex Loop

mindfulness (Tosey et al., 2011) and has been attributed to learning about the learning process through continuous reflection based on assumptions and values that motivate learning (Tosey et al., 2011). Looking for a more complete level of learning, triple-loop learning addresses not only content but also skills, competences, and infrastructures to support this learning. Learning is aided by "structural patterns (mental maps, facilitating structures, etc.). . . . the relationship between structure and behavior" (Georges and van Witteloostuijin, 1999: 440). Triple-loop learning could be perceived as a new form of sensemaking—that is, real-time sensemaking.[9]

Chaotic Loop

The final loop, the chaotic loop (Figure 5.6), is most similar to the clear loop. In the chaotic loop, however, one begins with acting rather than beginning with observing as in the clear and complicated loops. The feedback and feedforward mechanisms are also different because of the change in the initial starting condition. The act begins this stage so that one can observe interactions with

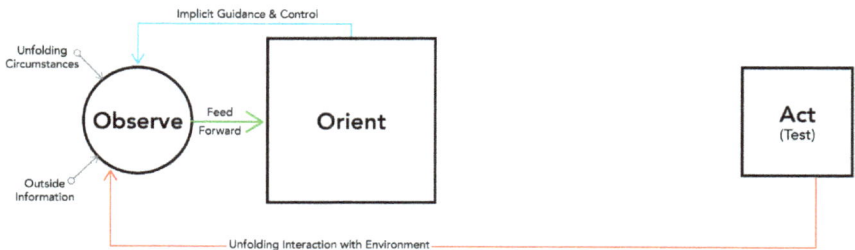

FIGURE 5.6.   Chaotic Loop

---

9   Real-time sensemaking was presented in conversations with Charlie Protzman who aided in reviewing and editing this book. Charlie conceptualized triple-loop learning with real-time sensemaking, so we decided to implement this terminology here.

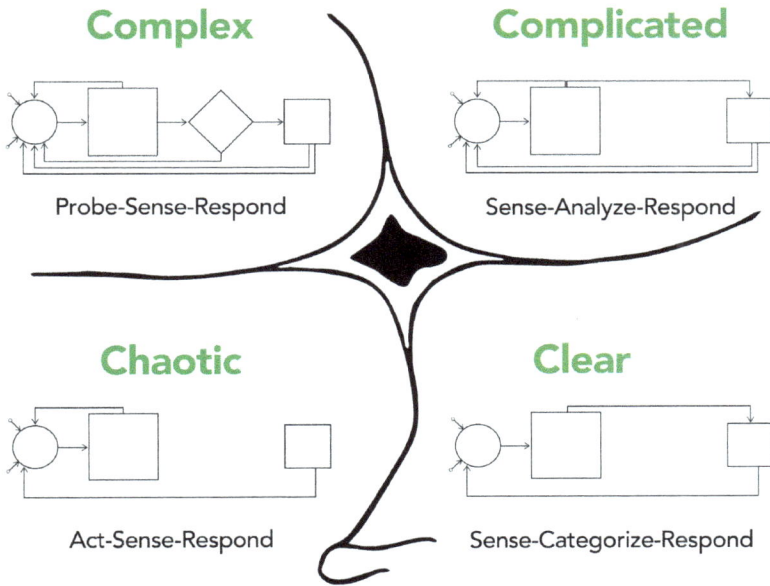

FIGURE 5.7.   Composite Cynefin/OODA Loop

environmental forces. Operating in this domain is recommended to be temporary, as in exploration or creativity, in which a quick experimental test is pursued to survey the landscape. Operating in the chaotic domain for any extended period of time is counterproductive because conditions change too rapidly and no clear determination can be made—that is, there are too many unknown-unknowns.

A Composite Model

By overlaying the Cynefin framework provided in Figure 5.1, with the different OODA loops shown in Figures 5.3 through 5.6, a composite representation of the Cynefin-OODA Loop is shown in Figure 5.7. Figure 5.7 highlights that different methods, techniques, and tools are required for each of the different domains. Operating in the complex domain, for example, requires more than just single-loop learning techniques; triple-loop learning is essential. This point is also highlighted in Figure 5.8, which summarizes different methods, techniques, and tools for each of the four domains. The examples used here (see Figure 5.8) include considerations for the type of team required for each domain, the type of knowledge base that needs to be utilized, the level of adaptability for each domain, the types of taskwork required, and the level of experimentation, experiential learning, and sensemaking. The main point projected is that each domain requires different methods, techniques, and tools to function effectively.

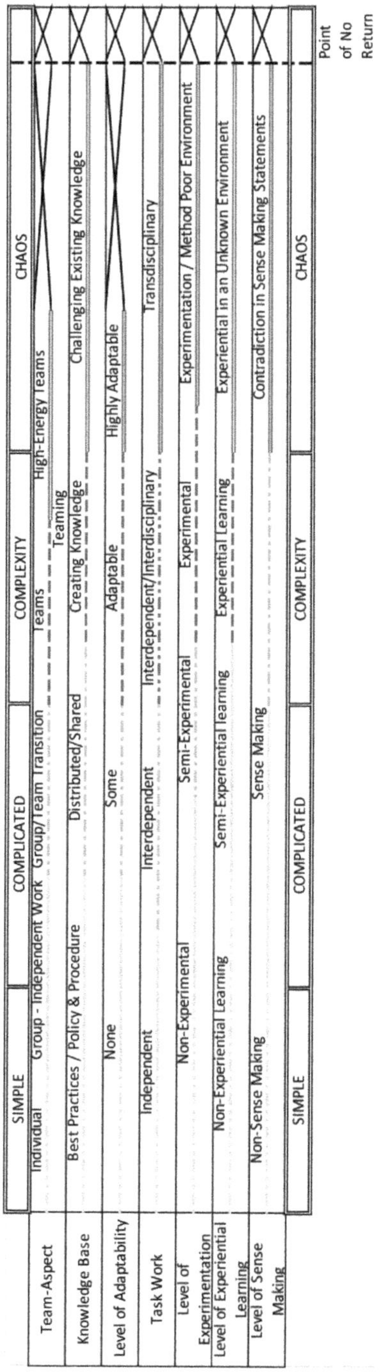

	SIMPLE	COMPLICATED	COMPLEXITY	CHAOS
Team-Aspect	Individual	Group - Independent Work / Group/Team Transition	Teams / Teaming / High-Energy Teams	
Knowledge Base	Best Practices / Policy & Procedure	Distributed/Shared	Creating Knowledge / Challenging Existing Knowledge	
Level of Adaptability	None	Some	Adaptable / Highly Adaptable	
Task Work	Independent	Interdependent	Interdependent/Interdisciplinary / Transdisciplinary	
Level of Experimentation	Non-Experimental	Semi-Experimental	Experimental / Experimentation / Method Poor Environment	
Level of Experiential Learning	Non-Experiential Learning	Semi-Experiential Learning	Experiential Learning / Experiential in an Unknown Environment	
Level of Sense Making	Non-Sense Making	Sense Making	Contradiction in Sense Making Statements	
	SIMPLE	COMPLICATED	COMPLEXITY	CHAOS

Point of No Return

FIGURE 5.8.   Different Methods, Tools, and Techniques for Each Domain

References

Boyd JR. (2018, March) *A discourse on winning and losing.* Maxwell Air Force Base, AL: Air University Press.

Chalmers AF. (2013) *What is this thing called science?* Queensland, Australia: University of Queensland Press.

Chia R. (2000) Discourse analysis as organizational analysis. *Organization* 7: 513–518.

Churchman WC. (1967) Wicked problems [editorial]. *Management Science* 14: B141–B142.

Cognitive Edge. (2010, August 24) *Summary article on origins of Cynefin.* Available at: https://cognitive-edge.com/articles/summary-article-on-cynefin-origins/.

Deming EW. (1994) *The new economics: For industry, government, education.* Cambridge, MA: MIT Press.

Deming EW. (2000) *Out of the crisis.* Cambridge, MA: MIT Press.

Georges RA and van Witteloostuijin A. (1999) Circular organizing and triple loop learning. *Journal of Organizational Change Management* 12: 439–454.

Godfrey-Smith P. (2003) *Theory and reality: An introduction to the philosophy of science.* Chicago, IL: University of Chicago Press.

Kauffman SA. (1993) *The origins of order: Self-organization and selection in evolution.* New York, NY: Oxford University Press.

Kurtz CF and Snowden DJ. (2003) The new dynamics of strategy: Sense-making in a complex and complicated world. *IBM Systems Journal* 42: 462–483.

Lewin K. (1951) *Field theory in social science.* New York, NY: Harper & Row.

Lewin R. (1992) *Complexity: Life at the edge of chaos.* New York, NY: Macmillan.

Liker JK. (2004) *The Toyota Way: 14 management principles from the world's greatest manufacturer.* New York, NY: McGraw-Hill.

Moen R and Norman C. (2009, September 17) Evolution of the PDCA Cycle. In: *Proceedings of the 7th ANQ Congress, Tokyo.*

Orlikowski WJ. (1996) Improvising organizational transformation overtime: A situated change perspective. *Information Systems Research* 7: 63–92.

Ravilious K. (2008, April 27) Perfect harmony. *The Guardian.*

Reinertsen DG. (2009) *The principles of product development flow: Second generation lean product development.* Redondo Beach, CA: Celeritas Publishing.

Roberts N. (2000) Wicked problems and network approaches to resolution. *International Public Management Review* 1: 1–19.

Romanelli E and Tushman ML. (1994) Organizational transformation as punctuated equilibrium: An empirical test. *Academy of Management Journal* 37: 1141–1166.

Snowden DJ. (2019, April 19) Liminal Cynefin & "control." Cognitive Edge. Available at: https://cognitive-edge.com/blog/liminal-cynefin-control/.

Snowden DJ. (2012a, July 25) *The origins of Cynefin: Part 3.* Cognitive Edge. Available at: http://old.cognitive-edge.com/wp-content/uploads/2010/08/The-Origins-of-Cynefin-Cognitive-Edge.pdf.

Snowden DJ. (2012b, July 25) *The origins of Cynefin: Part 5.* Cognitive Edge. Available at: http://old.cognitive-edge.com/wp-content/uploads/2010/08/The-Origins-of-Cynefin-Cognitive-Edge.pdf.

Snowden DJ and Boone ME. (2007) A leader's framework for decision making. *Harvard Business Review* 85: 68–76.

Tosey P, Visser M, and Saunders MN. (2011) The origins and conceptualizations of "triple-loop" learning: A critical review. *Management Learning* 43: 291–307.

Turner JR and Baker R. (2019) Complexity theory: An overview with potential applications for the social sciences. *Systems* 7: 23.

Weick KE. (2009) *Making sense of the organization: The impermanent organization.* West Sussex, UK: Wiley.

Weick KE and Quinn RE. (1999) Organizational change and development. *Annual Review of Psychology* 50: 361–386.

Woodside AG. (2017) *The complexity turn: Cultural, management, and marketing applications.* Cham, Switzerland: Springer.

Xiang W-N. (2013) Working with wicked problems in socio-ecological systems: Awareness, acceptance, and adaptation. *Landscape and Urban Planning* 110: 1–4.

CHAPTER 6

Systems Thinking Versus Complexity Thinking

Systems Thinking

Systems thinking relates to viewing organizations or institutions as systems, utilizing a holistic approach that includes the system, its subsystems, external influence, and interactions among the components of the system (Monat and Gannon, 2017). General systems thinking guidelines for solving problems include the following:

Step 1. Develop and articulate a problem statement.
Step 2. Identify and delimit the system.
Step 3. Identify the events and patterns.
Step 4. Discover the structures.
Step 5. Discover the mental models.
Step 6. Identify and address archetypes.
Step 7. Model (if appropriate).
Step 8. Determine the systemic root cause(s).
Step 9. Make recommendations.
Step 10. Assess improvement. (Monat and Gannon, 2017: 7)

These general systems thinking steps showcase the limitation of systems thinking when it comes to working with complexity or open systems. As previously discussed, complex or wicked problems are unclear and cannot be defined in a single, or a series, of problem statements. Open systems have no boundaries preventing one from delimiting the system. Although structures, mental models, archetypes, and a representative model could be developed, these will be only temporary as the conditions of the problem are constantly changing at varying speeds. Root cause, if determined, will be insufficient once the parameters of the complex or wicked problem change, which then leads to the cause and solution becoming irrelevant as interventions are put in place. For closed and semi-open systems, these general systems thinking guidelines will work. When operating in a complex environment and dealing with complex and wicked problems, different guidelines are in order.

Systems Thinking and Organizations

In presenting systems thinking as the foundation for learning organizations, Senge (2006) described systems thinking as a *shift of mind*. Systems thinking is about using systems theory to view problems. This shift of mind moves one from "seeing parts to seeing wholes" (Senge, 2006: 69). Applying systems thinking begins with understanding the Senge's *Laws of the Fifth Discipline*, which states the following:

Law 1: Today's problems come from yesterday's "solutions."
Law 2: The harder you push, the harder the system pushes back.
Law 3: Behavior grows better before it grows worse.
Law 4: The easy way out usually leads back in.
Law 5: The cure can be worse than the disease.
Law 6: Faster is slower.
Law 7: Cause and effect are not closely related in time and space.
Law 8: Small changes can produce big results—but the areas of highest leverage are often the least obvious.
Law 9: You can have your cake and eat it too—but not at once.
Law 10: Dividing an elephant in half does not produce two small elephants.
Law 11: There is no blame. (Senge, 2006: 57–67)

In viewing systems thinking, we can use an example taken from the *Columbia* space shuttle accident that occurred on February 1, 2003, with a loss of seven crew members. The National Aeronautics and Space Administration's (NASA) accident investigation report, titled the "Columbia Accident Investigation Board (CAIB) Report," identified both physical and organizational problems that contributed to the *Columbia* disaster. This systems thinking approach was evident in the following statement of the report: "The explanation is about system effects: how actions taken in one layer of NASA's organizational system impact other layers. History is not just a backdrop or a scene-setter. History is cause" (Columbia Accident Investigation Board, 2003: 195).

In looking at the first law, *Today's problems come from yesterday's "solution,"* the CAIB findings identified some of the temporal connections to actions or behaviors that had been taken in the past. Often, a historical record or account provides evidence of causation. The CAIB report stressed that the causal mechanism included "a wide range of historical and organizational issues, including political and budgetary considerations, compromises, and changing priorities over the life of the Space Shuttle Program" (Columbia Accident Investigation Board, 2003: 9).

The harder you push, the harder the system pushes back, the second law, is represented in the discovery of the red tape required to recommend any changes at NASA. These layers of "processes, boards, and panels . . . produced a false sense of confidence in the system" (Columbia Accident Investigation Board, 2003: 199). The buildup of these barriers, these layers of red tape, allowed confidence to build until it became counterproductive. This leads to an example of the third law, *Behavior grows better before it gets worst*.

The easy way out often leads back in, the fourth law, was evident in NASA's inability to dismiss the threat caused by the missing foam that acted as a heat shield for the space shuttle. The safety concerns surrounding the missing foam were discounted (i.e., the easy way out), with an understanding that "foam was a maintenance problem and a turnaround issue, not a safety-of-flight issue" (Columbia Accident Investigation Board, 2003: 200). Obviously, the failure to acknowledge the missing foam as a genuine safety risk (not acknowledging weak signals) forced NASA to answer to their decision-making processes after the failure, contributing to this easy way out making its way back in.

The cure can be worse than the disease, the fifth law, was evident in managements' responses to engineering. Managements' mind-set surrounding the missing foam was that it was safe, partially because of the time, effort, and expense required to actually fix the problem. Rather than show that the space shuttle was safe to fly, or to reenter the atmosphere, without the foam tiles, managements' mind-set was to have engineering "prove that it was unsafe to fly" (Columbia Accident Investigation Board, 2003: 201), reversing the role of engineering to fit managements' mind-set that the shuttle was safe. The cure, sending a second shuttle to repair the *Columbia* shuttle, was not discussed. This is an example of the cure being worse than the disease.

Faster is slower, the sixth law, refers to how, in some cases, a little knowledge can be a dangerous thing (Senge, 2006). In the *Columbia* investigation, this was prevalent after the accident as decisions were made by people who had status as opposed to those who had the expertise (Columbia Accident Investigation Board, 2003). In regards to the seventh law, *Cause and effect are not closely related in time and space*, one initially would have identified the missing foam tile as the cause of the accident. Upon closer inspection, however, the cause was multidimensional and systemic throughout NASA's culture.

The eighth law, *Small changes can produce big results—but the areas of highest leverage are often the least obvious*, is evident from the policy changes forced onto NASA. These policy changes also negatively affected the culture, structure, and safety of the organization, the whole. The culture, structure, and safety of NASA eventually led "NASA on its slippery slope toward *Challenger* and *Columbia*" (Columbia Accident Investigation Board, 2003: 197).

The ninth law, *You can have your cake and eat it too—but not at once*, is represented in managements' emphasis on safety even as it made cuts to key personnel.

This "kind of doublespeak by top administrators affects people's decisions and actions without them even realizing it" (Columbia Accident Investigation Board, 2003: 199). Reducing costs while maintaining a high level of safety could have been achieved; however, the manner in which this was actually attempted sent conflicting messages throughout the organization, hampering NASA's efforts to maintain safety goals.

The 10th law, *Dividing an elephant in half does not produce two small elephants,* is represented in the conflicting goals between those overseeing costs, scheduling, and safety. The CAIB report highlighted this conflict: "NASA had conflicting goals of cost, schedule, and safety. Safety lost out as the mandates of an 'operational system' increased the schedule pressure" (Columbia Accident Investigation Board, 2003: 200). Senge (2006) described this problem similar to organizational silos in which each department looks after their own policies and interests without considering how their policies interact and affect other departments, thus affecting the whole project or organization.

The 11th law, *There is no blame,* is indicative of the inability to freely speak one's mind during the trouble-shooting phase. In this case, management practiced confirmation bias by listening only to the information that they wanted to hear. Evidence of this was provided by the CAIB report: "Management did not listen to what their engineers were telling them" (Columbia Accident Investigation Board, 2003: 201). This confirming behavior, over time, resulted in a culture in which employees were intimidated to speak up: "Signals were overlooked, people were silenced, and useful information and dissenting views on technical issues did not surface at higher levels" (Columbia Accident Investigation Board, 2003: 201). Although there were plenty of instances in which a person, or a group of people, could be blamed for a small contributing factor that lead to the systemic failure, the blame goes to the overall process and the breakdowns that were allowed to occur to the system in each of the individual subsystems. The overall system is to blame rather than an individual or a group of individuals. Blaming an individual, or individuals, prevents one from getting to the root cause. Even if management had repositioned people into different positions or roles, they still would have been victims of the system, and the results would have been the same. The problem became systemic.

In the end, the unfortunate conclusions was that "relevant information that could have altered the course of events was available but was not presented" (Columbia Accident Investigation Board, 2003: 201). Although the investigation provided by the CAIB report was done in a post hoc manner, after the fact, it effectively showed how systemic breakdowns throughout the organization led to the unfortunate failure of the *Columbia* space shuttle. In real time, accounting for systemic issues is difficult and becomes extremely more so when multiple players are involved and when the size of the organization increases. The environment, at some point, can become a complex environment in which simple hierarchical structures and checks and balances simply will not work effectively. Weak signals (i.e., failing to acknowledge the missing foam as a genuine safety risk) are antecedents to larger systemic problems that often remain undetected. The CAIB report stressed this point: "While rules and procedures were essential for coordination, they had an unintended but negative effect. Allegiance to hierarchy and procedure had replaced deference to NASA engineers' technical expertise" (Columbia Accident Investigation Board, 2003: 200).

Aside from entering a complex environment, additional problems occur when encountering ill-defined or ill-structured problems. NASA identified that ill-structured problems (i.e., problems that are less visible) could be the most dangerous. NASA introduced the following recommended changes as an effort to "design systems that maximize the clarity of signals, amplify weak signals so they can be tracked, and account for missing signals" (Columbia Accident Investigation Board, 2003: 200). These changes include the following, beginning with leadership:

> *Leaders create culture. It is their responsibility to change it. . . . Changes in organizational structure should be made only with careful consideration of their effect on the system and their possible unintended consequences. . . . Strategies must increase the clarity, strength, and presence of signals that challenge assumptions about risk.* (Columbia Accident Investigation Board, 2003: 203; emphasis in original)

Systems Thinking Laws of Organizational Systems

Looking at organizations from a systems thinking perspective provides the fundamental laws of organizational systems:

1. Understanding performance requires documenting the inputs, processes, outputs, and customers that constitute a business.
2. Organization systems adapt or die.
3. When one component of an organization system optimizes, the organization often suboptimizes.
4. Pulling any lever in the system will have an effect on other parts of the system.
5. An organization behaves as a system, regardless of whether it is being managed as a system.
6. If you pit a good performer against a bad system, the system will win almost every time. (Rummler and Brache, 1995: 13)

These fundamental laws provide general guidelines for viewing an organization from the lens of systems thinking. First, a general understanding of systems theory is needed. Second, it must be understood that adaptation is required to remain sustainable. Third, the goal is not to optimize one system. Optimizing one of the systems, or subsystems, could require additional resources and energy that leads to suboptimization of other subsystems. A balancing act is required between optimization and systemic impact. Fourth, changes in one subsystem undoubtedly will have an effect on other subsystems and systems. Fifth, this is similar to Senge's (2006) last law (i.e., there is no blame); the system wins over the individual every time. Either the system works effectively, or it does not, but it is the system that produces the final outcome, not the individual. This last law has become the first place to begin when troubleshooting organizational problems: "Our experience strongly suggests that the Process Level is where the most substantive change usually needs to take place" (Rummler and Brache, 1995: 44).

Weak Signals

Part of the problem with current systems thinking guidelines or laws is that they do not account for weak signals. When weak signals are encountered, they often do not make enough of an impact to get noticed. Because the system does not change because of these weak signals, they are ignored. Unfortunately, it is these weak signals that should be amplified as they are often the initial indicators of a larger potential problem when not addressed. Maintaining the system and its output, either by qualitative measures or financial measure, is often the primary focus. Breakdowns in the system often are addressed in a reactive manner as opposed to a proactive manner, which also accounts for missing these weak signals.

Bounded Boundedness

Another problem comes when the system, the whole, becomes too bounded. As systems grow, controlling functions are put in place, which include placing additional boundaries around the system components, building silos. As more boundaries, or silos, are put in place, it becomes more difficult to communicate or to transfer information across these functional boundaries. As organizational problems occur, they often are addressed within the functional silo that is most affected, ignoring the larger systemic connections and interactions that occur between the organization's functional boundaries.

These boundaries can be cognitive in nature. Constraints are highlighted as being constructed: "Constraints are partly of one's own making and not simply objects to which one reacts" (Weick, 2009: 147). This leads to the need of externalizing each agent's understanding of which constraints are present, allowing weak signals to be detected. It is not necessarily that we need to remove, reduce, or manage these constraints as much as it is that we need to identify these perceived constraints.

This problem of mechanizing toward a technocratic society (von Bertalanffy, 1972), or organization, because of too many boundaries (silos), is similar to Bennett's description of *logical depth*. Logical depth refers to an organization and is defined as "the number of steps in the deductive or causal path connecting a thing with its plausible origin" (Bennett, 2019: 357). The farther one moves away from the originating causal mechanisms of a particular problem or issue, the more granularity surrounds the information about the problem or issue. As more boundaries or silos are placed around a system and its components, when problems do occur, they often are too granular to understand once they reach the decision makers. This makes the problem too difficult to resolve, if it is even recognized as a problem that needs to be addressed. This bounded boundedness not only weakens one's problem-solving capabilities but also buffers weak signals, making them nearly undetectable.

Anticomplexity Thinking

As more examples surface in which systems theory has been applied to open systems, controls have been placed upon these systems in an attempt to manage these systems, and more complex problems and issues begin to surface, it becomes apparent that systems thinking may not be appropriate for managing complexity. The following examples showcase some of the complications that arise when using systems theory and thinking practices to social systems when complexity arises.

Recent Examples of Systems Thinking Failures

General Motors

In 2014, a series of defective ignition switches that had been installed in nearly 2.6 million cars, General Motors (GM) had experienced a number of failures that led to approximately 13 known deaths (Bennet, 2014). In the end, GM paid a $35 million fine for its negligence (Bennet, 2014). The investigation found a cover-up within the ranks and determined that the problem was systemic: "A 'silo' culture in which managers in different departments failed to communicate safety concerns to one another or to senior executives" (Bennet, 2014: para. 12). A lack of sharing information within this siloed structure prevented managers from putting the pieces together. This siloed structure created more complexity and prevented GM from being able to adapt and move when emergencies did occur. These silos acted as boundaries around subsystems, creating an even more rigid structure for information to flow and eventually triggering one of the causes of the accidents. In addition to the rigid silos, weak signals, such as the different pieces of information that were never put together to tell the whole story, had never surfaced. This lack of awareness to weak signals was also a result of the cultural problems experienced at GM and of the siloed structure that had become frozen or standard.

Boeing 737 Max

At the time of this writing, Boeing is facing problems with its Boeing 737 Max. Boeing recently experienced two 737 Max plane crashes: "a Lion Air flight last October [2018] and an Ethiopian Airlines flight this March [2019]" (Robison, 2019: para. 4). Early reports are identifying similar problems that had been experienced with the *Challenger* and *Columbia* accidents. For example, it was reported that there was some "dismay over performance targets that risked sacrificing safety for profits" (Robison, 2019: para. 9). Boeing also experienced a deduction in its workforce with an influx of contractors to replace this deficit. These cuts, among others, came with a message from management that shareholders come first at Boeing, moving the customer, quality, and safety down the priority list.

The Boeing 737 Max uses what is called the Maneuvering Characteristics Augmentation System (MCAS), which incorporates software to address a mechanical deficit, keeping the plane airborne without stalling. This system was designed, in part, in response to the scaling of the original 737 to a larger version, the 737 Max. The MCAS proved to be a less expensive alternative to "modifying the airframe to accommodate the larger engines" (Travis, 2019: para. 22) as a fix to producing a "dynamically unstable airframe, the 737 Max" (Travis, 2019: So Boeing produced). These lessons, which we are only beginning to uncover, teach us about the real dangers that occur when addressing complexity (Travis, 2019), complexity that was self-generated by organizational agents. This issue is still being investigated and will be for some time. We draw no conclusions or make no blame. We only seek to highlight how the effects of complexity can lead to undesirable outcomes.

A Shift Toward Complexity Thinking

As these examples show (NASA's *Challenger* and *Columbia* space shuttles, GM's switches, Boeing's 737 Max), the risk of operating as usual (i.e., using systems thinking approaches to solve problems) when entering into newly undiscovered complex territories held devastating results: "Every increment, every increase in complexity, ultimately leads to decreasing rates of return and, finally, to negative returns" (Travis, 2019: Boeing in the process). New and different approaches are required when operating in complex environments and when addressing complex and wicked problems—methods and procedures that are different from those presented in systems thinking are needed. The Flow System (TFS) provides one such alternative when addressing complexity. The next section introduces the alternative to systems thinking, complexity thinking.

Complexity Thinking

Having the understanding that not everything can be predicted because of unknown-unknowns is the first step to achieving complexity thinking. Although teams and organizations operate with some level of certainty and prediction, it is important to also recognize that not everything is known or can be predicted. Take the number of interactions that take place between team members in a team-based environment as an example. It is safe to assume that as you increase the number of team members, you also increase the number of team member interactions. Team member interactions have been shown to be essential to measure when determining team or group effectiveness (Pentland, 2012). Team member interactions is an indicator of a team's communication pattern or information flow (from one team member to another). The greater the number of team members, the more interactions, and the more effective the team (at least this is the commonsense approach).

In reality, as you increase the number of team members in a small group, you also increase the complexity that takes place inside the group as well as external to the group. The total lines of communication can be represented by N(N-1)/2 (Brook's law), where "the added effort of communicating may fully counteract the division of the original task" (Brooks, 1995: 18). In Brook's law, N is the number of team members. Figure 6.1 provides a quick view of the increasing number of interactions that are required as the number of team members increase. In this example, a team with three members would require three lines of communication. As you increase the team size from 3 to 5, the number of interactions required would increase to 10, a team size of 7 requires 21 interactions, a team with 10 members require 45 interactions, and a team of 20 requires 190 interactions. If you were to add just one additional team member, say, from a team size of 6 to 7, the number of interactions increase from 15 to 21, requiring the new member to actively seek 6 new lines of communication.

It may not sound like much but increasing the number of interactions required of teams to successfully manage adds a level of complexity. In some instances, these interactions are not made, resulting in team members not having a say, or creating divisions within the team's structure because of the lack of communication. As teams become too big, you also find that some team members do

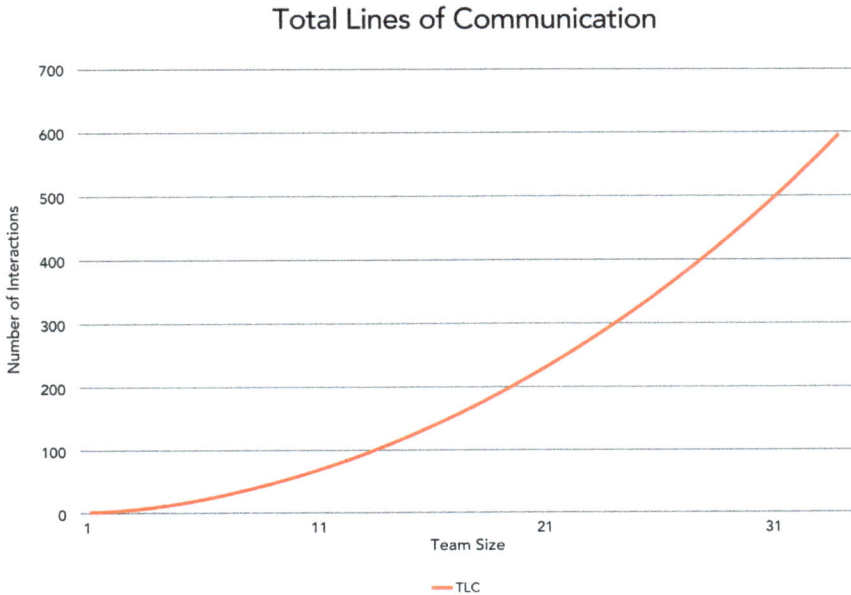

FIGURE 6.1. Total Lines of Communication

not have enough tasks to keep them busy, resulting in either disengaged or unmotivated team members, or you encounter social loafing[1] from a few team members. The increasing number of interactions because of large team sizes introduces complexity into the process. In one case, it was found to be a contributing factor that prevented team members from making accurate predictions (McChrystal et al., 2015) in fighting terrorism. This led McChrystal et al. to realize that "adaptability is more characteristic of small interactive teams than large top-down hierarchies" (2015: 113).

Although the previous discussion focused only on team interactions, complexity can be found in departments, organizations, institutions, governments, and local communities, as well as on a global scale. Similar to Brook's law previously discussed, Graicunas presented formulas that calculated different variations of an organization's span of control (Graicunas, 1937). Graicunas highlighted that "one of the surest sources of delay and confusion is to allow any superior to be directly responsible for the control of too many subordinates" (Graicunas, 1937: 183).

What makes an environment[2] a complex one is when the environment is fused with uncertainty and ambiguity with multiple options from which to choose. Complex environments involve multiple possible states that come and go over time, include different conditions across space, and are in a different state at different times (Godfrey-Smith, 1998). Complex environments do not maintain their current state, include heterogeneity, are disorganized and organized, and remain unpredictable. According to Godfrey-Smith, "Any environment will be complex in some respects but simple in others, and only some complexity properties will be relevant to any given organism. An environment may be simple at one level or time-scale but complex at another" (Godfrey-Smith, 1998: 25).

According to Godfrey-Smith (1998), when addressing complex environments, one must answer the following two questions:

1. At what state is the current environment?
2. How much variability is in the current environment?

The first question makes organizations cognizant of whether or not they are in a complex, complicated or simple environment. The latter indicates the frequency at which the conditions in the environment change. If the environment is not

---

1   Social loafing can be defined as a reduction in individual effort (Forsyth, 2014). When teams or groups get too large, social loafing becomes a consideration. Many times, too much time is spent on coordinating activities compared with actual taskwork; here, social loafing becomes a problem because not everyone can be involved in coordinating activities. Social loafing is not necessarily a characteristic of any one individual as much as it is a characteristic of large groups.

2   Although environment is used in this discussion, this term could be changed easily with group, organization, or country. Environment refers to situational and contextual characteristics in an unbounded state.

complex, with little to no variability, then systems thinking methods should work fine. When complexity is present with high variability, however, complexity thinking is in order.

Distinguishing between complicated and complexity is also in order. Complicated involves dealing with linear systems that can be accurately evaluated (e.g., airliner, cargo ship), but when the system becomes nonlinear as a result of intricate interactions among components, the system becomes complex (e.g., using software to correct the Boeing 737 Max, autonomous cargo ships) (Kruger et al., 2019).

This then begins the transition between systems thinking and complexity thinking. Organizations are both complicated and complex, and often operate simultaneously in both domains.[3] In an organization's habitat (Cynefin's translation), the organization and its agents are influenced by factors in both the complicated and complex domains as they try to make sense of the unknowns. The ability to know the domain (Cynefin framework) in which one is operating at any given time is essential before determining the course of action to take moving forward. The action taken when addressing simple or complicated problems (systems thinking) is different from the action required to address complex problems (complexity thinking). Although the goal is to utilize complexity thinking to move complex problems into complicated waters that are easier to navigate, doing so requires a different type of thinking, complexity thinking.

One critical difference between complicated (systems thinking) and complex (complexity thinking) settings or environments can be found in the following description provided by Richardson: "Basically, *absolute knowledge about the parts that make up a system and their interactions provides us with very little understanding indeed regarding how that system will behave overall*" (Richardson, 2008: 14–15; emphasis added). Utilizing general systems thinking is insufficient in understanding or predicting how the whole system will behave based on information from the subsystems or components of the whole. By dividing a system, or problem, into parts, one is destroying what it seeks to understand (Kruger et al., 2019). This point has been highlighted by other scholars and practitioners: "Attempts to control complex systems by using the kind of mechanical, reductionist thinking championed by thinkers from Newton to Taylor-breaking everything down into component parts, or optimizing individual elements-tend to be pointless at best or destructive at worst" (McChrystal et al., 2015: 69).

Transitioning to complexity thinking has been understood, abstractly, for some time:

> As the world becomes more complex and interdependent, the ability to think systematically, to analyze fields of forces and understand their joint causal effects on each other, and to abandon simple linear causal logic in favor of complex mental models will become more critical to learning. The learning leader must believe that the world is intrinsically complex, non-linear, interconnected, and "overdetermined" in the sense that most things are multiply caused. (Schein, 2010: 371)

---

3   We refer to the domains as different influences from one's habitat (Cynefin's translation).

What is needed, and what is presented here, is a foundation to begin this transition to complexity thinking. Complexity thinking can be defined as the methods and tools to understand complexity with a worldview of seeing complex adaptive systems (CASs) as opposed to simple systems and subsystems. The steps toward complexity thinking, along with additional explanations for each step, including CASs, are addressed in the next section.

The Steps to Complexity Thinking

The road map presented here is theoretical, supported by empirical research and practice. Tools provided to help implement complexity thinking in practice are presented. As this field of study continues to grow, the tools to support complexity thinking will flourish and become readily available, especially as technologies such as network analysis, big data analysis, artificial intelligence, and machine learning intensify. Theoretically, it is believed that the road map toward complexity thinking that is presented here is a work in progress. As with any theory, "a theory is scientific to the degree to which it is testable" (Popper, 1985: 123). Testing this theory is essential along with the iterative process of modifying the theory and continually testing it until, one day, we have a potential theoretical model that could be used in a number of different contextual settings. Knowing that no one theory will apply to every situation or setting, theories must be modified to meet specific situations. The theory presented here is the starting point to begin testing the components of complexity thinking:

> We choose the theory which best holds its own in competition with other theories; the one which, by natural selection, proves itself the fittest to survive. This will be the one which not only has hitherto stood up to the severest tests, but the one which is also testable in the most rigorous way. A theory is a tool which we test by applying it, and which we judge as to its fitness by the results of its applications. (Popper, 2002: 91)

Ultimately, this theory will be tested and improved on to the point at which it has utility that meets our needs. Pragmatically, this theory should be submitted to testing to determine its utility for your own needs. In the long run, tested theories can be combined to make one grand theory, but this grand complexity thinking theoretical model is a long way away. Currently, we begin this theory testing stage with the model presented here. This model can be summarized in two steps:

- *Step 1* involves understanding the characteristics of complex systems.
- *Step 2* involves having a worldview or perspective that systems, entities, and events are CASs.

These general steps provide guidance for managing complexity and addressing complex problems. We must reiterate, however, that operating in complexity is an exploratory process in which no one person can fully understand the whole (Kruger et al., 2019). Complexity presents problems that are multidimensional, operate in different spatial and temporal zones, and are open with constant influences from environmental forces. These problems make it difficult or nearly impossible to understand complexity in its entirety. Operating in complexity requires exploratory processes to understand the whole, but "knowledge is always provisional" (Kruger et al., 2019: 15). Complexity thinking, in part, focuses on what cannot be explained as opposed to reductionistic methods that focus on what can be explained (Raisio and Lundstrom, 2017). This type of focus requires one to accept a certain level of uncertainty when making decisions to act: "We have to make decisions without having a model or a method that can predict the exact outcome of those decisions" (Cilliers, 2000: 28). This requires, then, having confidence in the various methods and techniques used when addressing complexity. The evidence-based steps presented here should provide some level of confidence. Each case is contextual, however, and requires adjustments, and thus a level of theory development and theory testing is also necessary.

Step 1. Characteristics of Complex Systems

Understanding the characteristics of complex systems is required before being able to transition to complexity thinking. Complex systems have the following characteristics in common:

1. Complex systems consist of a large number of elements that in themselves can be simple.
2. The elements interact dynamically by exchanging energy or information. These interactions are rich. Even if specific elements only interact with a few others, the effects of these interactions are propagated throughout the system. The interactions are nonlinear.
3. There are many direct and indirect feedback loops.
4. Complex systems are open systems—they exchange energy or information with their environment-and operate at conditions far from equilibrium.
5. Complex systems have memory, not located at a specific place, but distributed throughout the system. Any complex system thus has a history, and the history is of cardinal importance to the behavior of the system.
6. The behavior of the system is determined by the nature of the interactions, not by what is contained within the components. Since the interactions are rich, dynamic, fed back, and, above all, nonlinear, the behavior of the system as a whole cannot be predicted from an inspection of its components. The notion of "emergence" is used to describe this aspect. The presence of emergent properties does not provide an argument against causality, only against deterministic forms of prediction.
7. Complex systems are adaptive. They can (re)organize their internal structure without the intervention of an external agent. (Cilliers, 2000: 24)

Step 2. Complex Adaptive Systems

The general idea behind CASs is that the whole is more complex than its parts (Holland, 1998) and also more complicated and meaningful than the aggregate of its parts. Identifying the rules of emergence is essential before one can fully understand a CAS (Holland, 1998). Emergence is often recognized in the form of patterns—that is, new and unexpected patterns that *emerge* from a CAS. Tools capable of highlighting such patterns (e.g., network analysis, graph theory techniques) are necessary for indicating when emergence occurs as well as capturing the conditions that led up to the emergence so that the conditions could be replicated, if desired. Modeling and simulation techniques also can be used to capture the conditions that led to the desired behavior or outcome that resulted from the emergent event. Modeling techniques can become an essential tool for understanding emergence from a CAS. The following example highlights the benefits that modeling can provide when trying to better understand emergent behaviors:

> Accordingly, we can accumulate examples of emergence that are quite different one from another. Because of these differences, we can use a general setting to compare the examples and throw out properties that are incidental. This process enhances our chances of discovering, and controlling, the essential conditions for emergence. (Holland, 1998: 14–15)

CASs are of particular interest when viewing social systems (e.g., teams, organizations, governments). Social systems have the characteristics of being "diverse, nonlinear, consisting of multiple interactive, interdependent, and interconnected sub-elements. They are adaptive and self-organizing, tending toward ever-greater complexity operating at the 'edge of chaos,' and therefore in a constant state of innovation and dynamic equilibrium" (Waddock et al., 2015: 996). Pushed too far, CAS potentially can reach the level of being chaotic, reaching a "major state change or . . . collapse" (Waddock et al., 2015: 996).

An example of a social system being viewed as a CAS is found in Schumpeter's theory of economic development (Metcalfe, 2014). From this theory, capitalism was described as a CAS from the following description:

> Capitalism . . . is an adaptive system that responds to and induces the emergence of novelty from within, so that every pattern of order contains within it the unknowable seeds of its own destruction, it is a scheme in which unpredictable human creativity is the *sine qua non*[4] of change to prevailing structures. In capitalism, every position is open to potential challenge, it is an open system in which the ordering of economic society is ever changing and it never is and never can be at rest. (Metcalfe, 2014: 225)

This description of capitalism clearly highlights CAS in which capitalism in this example could be replaced with most constructs or entities. For example, one could replace capitalism with organization, project management, research and

---

4   *Sine qua non* refers to an essential action, condition, or ingredient.

development, or team, to name but a few. The key components of CAS are highlighted in this description and are described in more detail in the next section.

Complexity views CASs as being open systems composed of the following key characteristics: path dependency, system history, nonlinearity, emergence, irreducibility, adaptability, operation between order and chaos, and self-organization.

CAS Characteristics

Multiple descriptions of CAS can be found in the literature, but one notable study conducted a review of various characteristics of CAS (Turner and Baker, 2019a). From this review, the following tenets of CAS were identified as the most common characteristics found in the literature: path dependency, system history, nonlinearity, emergence, irreducibility, adaptability, operation between order and chaos, self-organization. Each of these tenets were defined in a follow-up research study by Turner and Baker (2019b). The definitions of each tenet that make up CAS are discussed next.

Path Dependency

The tenet of path dependency refers to systems being sensitive to their initial conditions. Each individual system or organism begins at a different time and place compared with other similar systems or organisms. This results in systems behaving differently to similar forces (Lindberg and Schneider, 2013), in part because of their initial conditions and histories.

In looking at agency theory, for example, an agent (individual) is afforded free will to act within the constraints and boundaries of the organization on behalf of the organization. The agent also has a personal set of motivations and needs expected of the organization. The agents' needs are partially based on their background, education, socioeconomic status, and culture. These needs also vary depending on one's position of power. Agents' needs at lower levels within an organization will be different from the needs of agents at higher organizational levels. Agency theory views the interactions between the different agents within an organization and their different needs and motivations. "Agency theory is based on cooperating parties seeking to have economic benefits, either personally or for the organization in which the social relationship is contained" (Baker, 2019: 2).

This relationship between agents partially depends on each agents' needs and motivations; this relationship is a path-dependent product. The literature highlights a few examples: agents with higher power tend to be more risk averse, passing the risk to lower-level agents (Baker, 2019). Other perspectives include conflict between agents because each agent is a stakeholder in the organization with different needs and motivations (Baker, 2019). These stakeholder positions could be perceived as being path-dependent positions as well. These examples highlight the importance of viewing agent relationships, events, and dynamics as being path dependent. Aside from viewing just agent systems as being path dependent, it is critical for leadership and for agents to view interactions as being path dependent, which is the beginning of the transition to viewing CAS.

System History

In addition to a system being a component of its initial conditions, path dependency, systems also are a component of their initial starting point. Each system's history could influence how it responds to changes (Boal and Schultz, 2007). Organizations, or other CASs, are components of their history and leadership must be capable of externalizing this history: "Strategic leaders play a central role in the organization's capacity to learn from its past, adapt to its present, and create its future" (Boal and Schultz, 2007: 411). Leadership must provide a bridge between the past, present, and the future (see strategic leadership) (Boal and Hooijberg, 2000) to present a road map for organizations to evolve and to foster a culture of creativity (Boal and Schultz, 2007).

In extending the previous agent theory example, as multiple agents interact over time, they begin to develop their shared knowledge and mental models. Individual agents draw on past events to build their models to react to future events: "Members draw on knowledge and experience from the past to create a capacity for action in the present, while also considering how the present fits with or accomplishes members' sense of the organization's future" (Boal and Schultz, 2007: 18). It is essential for leadership and managers to understand how these shared models developed and what conditions led to these shared models. The history of how these shared models developed is critical before one can understand what these shared models are and how one can begin to change these shared models.

Nonlinearity

The component of nonlinearity refers to the fact that systems react to external perturbations disproportionately and that the outcomes from nonlinear systems are different from those of simple systems (Lindberg and Schneider, 2013; Luoma, 2006). Related to the concept of dissipative structures, nonlinearity results in inconsistent or unstable outputs as opposed to reaching a state of equilibrium (Schneider and Somers, 2006). In addition to being inconsistent, each iteration produces a subsequent different outcome that add to the level of complexity: "At each point, the structure moves to a new and generally higher level of complexity that is qualitatively and quantitatively different from previous states" (Schneider and Somers, 2006: 354). Emergence, the next characteristic, occurs from the interactions between agents during each of these subsequent states.

Linearity is comfortable. Linear systems are easy to formulate, which is one reason why most sciences tend to live in a linear world, with linear methods and tools. Linear equations produce straight lines on a graph, thus providing some level of predictability. As an example, look at the number of research studies that utilize multiple regression analysis technique. The basic regression equation is of the form of a straight line ($y = mx + b$). In contrast, nonlinear systems are unsolvable and cannot be aggregated (Gleick, 2008), making them more difficult to provide any level of prediction.

A simple example of a nonlinear relationship can be found when looking at incentivizing workers. As one increases the level of pay for work, one often expects the level of motivation to complete this work to increase with pay. This linear relationship, increased pay resulting in increasing motivation, works up to a point. As some point, possibly after an employee has reached a level of social comfort and professional confidence, this linear relationship does not hold true. Increasing pay does not result in an additional increase in motivation. Perhaps one's motivation already has maxed out.

Other explanations claim that people are motivated by finding purpose in the work that they do more so than in the level of incentives received for the work. In his book *Drive*, Dan Pink has shown that motivation comes from one obtaining autonomy, a level of mastery, and purpose in what they do (Pink, 2009). Other research has shown that "persistence of an individual's state of engagement over time stems from the meaning that work or job activities have for an individual" (Shuck et al., 2017: 267). In these simple examples, the relationship between pay and motivation shifted to a new relationship between motivation and meaning of work. This phase shift is sometimes referred to as a bifurcation, in which simple relationships shift to new and different relationships. This concept is also similar to fractal distinctions, which are defined as patterns that repeat within themselves (Abbott, 2001). Fractal distinction, or repeated patterns, can occur within the same system or organization.

The initial relationship between pay and motivation becomes more complex once purposeful and meaningful work enters into the picture, and it becomes even more complex when time constraints and performance are included. The initial relationship presented here was a simple reductionistic relationship. As one moves outward to view this dynamic, by removing barriers, one begins to see the relationships between meaningful works. The farther one moves outward, and the more barriers that are removed, it becomes easier to identify the patterns that lead to the desired outcome, in this case, individual performance.

The initial linear and comfortable relationship quickly became a complex set of relationships, but they remain manageable. This type of shift in thinking, viewing beyond linear models and removing traditional barriers, is required to view CAS and for complexity thinking.

Emergence

The idea of emergence creates the perception that much comes from little (Holland, 1998). Simple laws of emergence include the following: "(a) the component mechanisms interact without central control, and (b) the possibilities for emergence increase rapidly as the flexibility of the interactions increases" (Holland, 1998: 7). Each system's internal dynamics affect its ability to change in a manner that might be quite different from other systems (Lindberg and Schneider, 2013). Emergence occurs through self-organizing systems in which individual agents interact, learn, and adapt to environmental forces to the point at which a new or modified system is created to meet the new demands of the environment. This new emerging system is unplanned, is not predictable, and occurs at the right time when the conditions are right.

Some examples of simple emergence are provided by Waldrop:

> Flying boids (and real birds) adapt to the actions of their neighbors, thereby becoming a flock. Organisms cooperate and compete in a dance of coevolution, thereby becoming an exquisitely tuned ecosystem. Atoms search for a minimum energy state by forming chemical bonds with each other, thereby becoming emergent structures known as molecules. Human beings try to satisfy their material needs by buying, selling, and trading with each other, thereby creating an emergent structure known as a market. Humans likewise interact with each other to satisfy less quantifiable goals, thereby forming families, religions, and cultures. Somehow, by constantly seeking mutual accommodation and self-consistency, groups of agents manage to transcend themselves and become something more. (Waldrop, 1992: 288–289)

Emergence can be identified using connectionist methods as opposed to reductionistic methods. Network analysis tools and computer modeling have been shown to be extremely successful in identifying frameworks for emergent patterns. These frameworks provide opportunities to capture the essence of emergence (Waldrop, 1992) while eliminating noise. To date, emergence has been attributed to the interconnections between nodes or agents: "the lesson is clear: the power really does lie in the connections" (Waldrop, 1992: 291). Ideally, once an emergent pattern has been identified and an acceptable framework for it has been derived, desired emergent behaviors are reinforced by fostering the interconnections in the framework. In contrast, undesired emergent behaviors are eliminated by preventing certain interconnections in the framework from occurring. Managing the interconnections is essential for promoting desirable emergent behaviors once an acceptable emergence framework has been identified. Ultimately, leadership plays an essential role in fostering emergence: "leadership should work toward achieving desirable order without harming emergent processes" (Lindberg and Schneider, 2013: 232).

Irreducibility

Irreversible process transformations cannot be reduced back to their original state (Borzillo and Kaminska-Labbe, 2011). Take, for example, learning. Once someone learns something, that person cannot purposively unlearn the new knowledge nor stop knowing how this new knowledge becomes associated with, or influences, existing knowledge. This example becomes more complicated when you look at small groups.

When looking at team learning as an example, team learning is not just the aggregate of the individual learning of each team member. Team learning also becomes a component of the interactions that take place between team members and the shared knowledge that develops between team members. In this example, team learning is greater than each individual team member's learning, and it becomes significantly different as it becomes more of a shared construct than an individual construct. In essence, the whole becomes greater than, and different from, the sum of its parts.

Adaptability

The tenet of a CAS being adaptive relates to having the capability to operate in systems that are simultaneously ordered and disordered; adaptive systems are more resilient (Borzillo and Kaminska-Labbe, 2011). Having the capability to process information about the environment is critical to adapting to environmental forces. Developing mental models about the frozen components of a system while, at the same time, practicing sensemaking techniques to understand the unfrozen components, and synthesizing the two into a whole, results in being adaptive. As the number of unfrozen components increase, it becomes more difficult to know what is going on; hence, more complexity is present. Equally so, as the total number of components increases, frozen and unfrozen components, it becomes more difficult to process complete sets of information, thus leading to more complexity.

Viewing adaptability as a component of information processing, Francisco Ayala provided the following when talking about evolutionary progress: "The ability to obtain and process information about the environment, and to react accordingly, is an important adaptation because it allows that organism to seek out suitable environments and resources and to avoid unsuitable ones" (as cited in Lewin, 1992: 138). Being adaptive results in having the capabilities to navigate toward more suitable environments or conditions rather than toward undesirable conditions. Leadership and management should facilitate adaptability into their lower-level units (teams and departments) while providing the appropriate training and resources to navigate complex environments, for favorable conditions and emergence to surface.

Operation Between Order and Chaos

Adaptive tension emerges from the energy differential between the system and its environment. This energy is transferred from the system to the external forces, creating the concept of a system operating on the "edge of order" or on the "edge of chaos" (Borzillo and Kaminska-Labbe, 2011). This concept of providing a balance between order and chaos is an important one to keep in mind at all times. This balance will result in one looking at how a system behaves as opposed to looking at how a system is made (Waldrop, 1992). Once the behavior of a system is in focus, one can identify the extremes of order and chaos (Waldrop, 1992).

Some system states are frozen, whereas other system states will be unfrozen. The idea is to obtain a healthy balance between the two rather than existing in a state that is primarily too frozen (order) or that is too unfrozen (chaos). Failing to provide this balance could result in a state that is primarily frozen, creating a system that is too rigid, inflexible, and unable to adapt. The former Soviet Union is a clear example in which a totalitarian centralized approach to organizing society was too "stagnant, too locked in, too rigidly controlled to survive" (Waldrop, 1992: 294). On the opposite end, "anarchy doesn't work very well, either" (Waldrop, 1992: 294).   *polarity*

What is needed is a balance between the two—that is, order and chaos—bottom-up and top-down processes. Organizations, for example, must adhere to specific rules and regulations, top-down, while also providing opportunities for creativity and experimentation, bottom-up, to flourish.

Take the Andon cord that is available in manufacturing facilities and is part of the Toyota Production System (TPS). The TPS provides a type of structure with top-down rules and procedures to follow during the manufacturing process. But it also provides bottom-up tools such as the Andon cord. If an employee on the assembly line identifies a problem, they have the opportunity to pull the Andon cord, so that they or others can look into a situation to see whether a real problem exists. Employees feel free to pull the Andon cord whenever they see a problem because they know they will not be penalized for doing so. The TPS provides structure but also provides a level of flexibility that allows bottom-up input and feedback to the overall process, balancing between top-down and bottom-up processes. Providing too much control, as in removing employees with the capability of providing input, likely will destroy the creativity capacity of any organization. Not providing enough support for bottom-up processes could prevent opportunities for experimentation to take place. Managing this balance is essential for any organism or entity (organization, department, team).

Self-Organization

Systems are composed of interdependency, interactions of its parts, and diversity (Borzillo and Kaminska-Labbe, 2011). Within CAS, self-organization occurs "through the dynamics, interactions and feedbacks of heterogeneous components" (Varga, 2014: 12). This characteristic is essential in the development of many other characteristics of CAS. For example, for emergence to occur, the system components must be self-organizing. In addition, for a system to be adaptable, it must have the characteristic of self-organization. This characteristic of self-organization is probably the interlinking characteristic of CAS, supporting each of the other characteristics.

Conclusion

As highlighted by Richardson: "If there are limits to what we can know, then there are of course limits to what we can achieve in a pre-determined, planned way" (Richardson, 2008: 13). Complexity thinking aids in accepting the unknowable and provides tools and techniques to help navigate complexity. The steps of complexity thinking presented in this chapter aid leaders, managers, and individuals to change their current thinking processes when dealing with complex problems and environments. Perhaps, at the very least, one could hope to avoid making existing environments complex environments by understanding the general rules of complexity thinking. Complexity thinking is just the first concept of three concepts presented in TFS (complexity thinking, distributed leadership, team science). Leadership and utilizing the benefits of teams are essential in navigating complexity and are presented in the next two chapters.

As a general guide for leaders and managers to acknowledge complexity, Cilliers proposed the following concepts about which leaders and managers should be mindful:

- Although systems which filter data enable us to deal better with large amounts of it, it should be remembered that filtering is a form of compression. We should never trust a filter too much.
- Consequently, when we talk of mechanized knowledge management systems, we can (at present?) only use the word "knowledge" in a very lean sense. There may be wonderful things to come, but a present I do not know of existing computational systems that can in any way be seen as producing "knowledge." Real breakthroughs are still required before we will have systems that can be distinguished in a fundamental way from database management. Good data management is tremendously valuable, but cannot be substitute for the interpretation of data.
- Since human capabilities in dealing with complex issues are also far from perfect, interpretation is never a merely mechanical process, but one that involves decisions and values. This implies a normative dimension to the "management" of knowledge. Computational systems which assist in knowledge management will not let us escape from this normativity. Interpretation implies a reduction in complexity. The responsibility for the effects of this reduction cannot be shifted away onto a machine.
- The importance of context and history means that there is no substitute for experience. Although different generations will probably place the emphasis differently, the tension between innovation and experience will remain important.
- We should manage complex systems with respect for the diversity they contain. Knowledge emerges from form the interaction between many different components. Thus, if we can maintain a rich diversity, the resources in the system will be richer. . . . Calling for "excess diversity." A system should not only have the "requisite variety" it needs to cope with its environment (Ashby's law). It should have more variety. Excess diversity in the system allows the system to cope with novel features in the environment. What is more, if a system has more diversity than it needs in order to merely cope with its environment, it can experiment internally with alternative possibilities. These creative capabilities should not "managed out." (Cilliers, 2014: 6–7)

References

Abbott AD. (2001) *Chaos of disciplines*. Chicago: University of Chicago Press.

Baker R. (2019) The agency of the principal-agent relationship: An opportunity for HRD. *Advances in Developing Human Resources*. Advance online publication: https://doi.org/10.1177/1523422319851274.

Bennet J. (2014, June 4) GM recall proble to blame cultural failings. *Wall Street Journal*.

Bennett CH. (2019) Dissipation, information, computational complexity and the definition of organization. In: Pines D (ed) *Emerging syntheses in science: Proceedings of the founding workshops of the Santa Fe Institute*. Santa Fe, NM: Santa Fe Institute Press, 357–383.

Boal KB and Hooijberg R. (2000) Strategic leadership research: Moving on. *The Leadership Quarterly* 11: 515–549.

Boal KB and Schultz PL. (2007) Storytelling, time, and evolution: The role of strategic leadership in complex adaptive systems. *The Leadership Quarterly* 18: 411–428.

Borzillo S and Kaminska-Labbe R. (2011) Unravelling the dynamics of knowledge creation in communities of practice though complexity theory lenses. *Knowledge Management Research and Practice* 9: 353–366.

Brooks FP. (1995) *The mythical man-month: Essays on software engineering*. Boston, MA: Addison-Wesley.

Cilliers P. (2000) What can we learn from a theory of complexity? *Emergence: Complexity and Organization* 2: 23–33.

Cilliers P. (2014) Knowledge, complexity and understanding. In: Strathern M and McGlade J (eds) *The social face of complexity science*. Litchfield Park, AZ: Emergent Publications, 1–8.

Columbia Accident Investigation Board. (2003, August) Report of the Columbia Accident Investigation Board. Washington, DC: NASA.

Forsyth DR. (2014) *Group dynamics*. Belmont, CA: Cengage Learning.

Gleick J. (2008) *Chaos: Making a new science*. New York, NY: Penguin Books.

Godfrey-Smith P. (1998) *Complexity and the function of mind in nature*. New York, NY: Cambridge University Press.

Graicunas VA. (1937) Relationship in organization. In: Gulick L and Urwick LF (eds) *Papers on the scinece of administration*. New York, NY: Columbia University, 181–187.

Holland JH. (1998) *Emergence: From chaos to order*. Reading, MA: Helix Books.

Kruger H, Verhoef A, and Preiser R. (2019) The epistemological implications of critical complexity thinking for operational research. *Systems* 7: 20.

Lewin R. (1992) *Complexity: Life at the edge of chaos*. New York, NY: Macmillan.

Lindberg C and Schneider M. (2013) Combating infections at Maine Medical Center: Insights into complexity-informed leadership from positive deviance. *Leadership* 9: 229–253.

Luoma M. (2006) A play of four arenas: How complexity can serve management development. *Management Learning* 37: 101–123.

McChrystal S, Collins T, Silverman D, et al. (2015) *Team of teams: New rules of engagement for a complex world*. New York, NY: Penguin.

Metcalfe S. (2014) Complexity and economic evolution from a Schumpeterian perspective. In: Strathern M and McGlade J (eds) *The social face of complexity science*. Litchfield Park, AZ: Emergent Publications, 223–244.

Monat JP and Gannon TF. (2017) *Using systems thinking to solve real-world problems.* Worcester, MA: College Publications.

Pentland A. (2012) The new science of building great teams. *Harvard Business Review* 90: 60–70.

Pink D. (2009) *Drive: The surprising truth about what motivates us.* New York, NY: Riverbend Books.

Popper K. (2002) *Popper: The logic of scientific discovery.* New York, NY: Routledge Classics.

Popper KR. (1985) *Popper selections.* Princeton, NJ: Princeton University Press.

Raisio H and Lundstrom N. (2017) Managing chaos: Lessons from movies on chaos theory. *Administration and Society* 49: 296–315.

Richardson KA. (2008) Managing complex organizations: Complexity thinking and the science and art of management. *Emergence: Complexity and Organization* 10: 13–26.

Robison P. (2019, May 8) Former Boeing engineers say relentless cost-cutting sacrificed safety. *Bloomberg Businessweek.*

Rummler GA and Brache AP. (1995) *Improving performance: How to manage the white space on the organization chart.* San Francisco, CA: Wiley.

Schein EH. (2010) *Organizational culture and leadership.* San Francisco, CA: Wiley.

Schneider M and Somers M. (2006) Organizations as complex adaptive systems: Implications of complexity theory for leadership research. *The Leadership Quarterly* 17: 351–365.

Senge PM. (2006) *The fifth discipline: The art and practice of the learning organization.* New York, NY: Currency Doubleday.

Shuck B, Osam K, Zigarmi D, et al. (2017) Definitional and conceptual muddling: Identifying the positionality of employee engagement and defining the construct. *Human Resource Development Review* 16: 263–293.

Travis G. (2019, April 18) How the Boeing 737 max disaster looks to a software developer. *IEEE Spectrum.*

Turner JR and Baker R. (2019a) Complexity theory: An overview with potential applications for the social sciences. *Systems* 7: 23.

Turner JR and Baker R. (2019b) Creativity and innovative processes as complex adaptive systems: A multilevel theory. Manuscript submitted for publication.

Varga L. (2014) Complexity science: The integrator. In: Srathern M and McGlade J (eds) *The social face of complexity science.* Litchfield Park, AZ: Emergent Publications, 11–25.

von Bertalanffy L. (1972) The history and status of general systems theory. *Academy of Management Journal* 15: 407–426.

Waddock S, Meszoely GM, Waddell S, et al. (2015) The complexity of wicked problems in large scale change. *Journal of Organizational Change Management* 28: 993–1012.

Waldrop MM. (1992) *Complexity: The emerging science at the edge of order and chaos.* New York, NY: Simon & Schuster Paperbacks.

Weick KE. (2009) *Making sense of the organization: The impermanent organization.* West Sussex, UK: Wiley.

PART III

Distributed Leadership

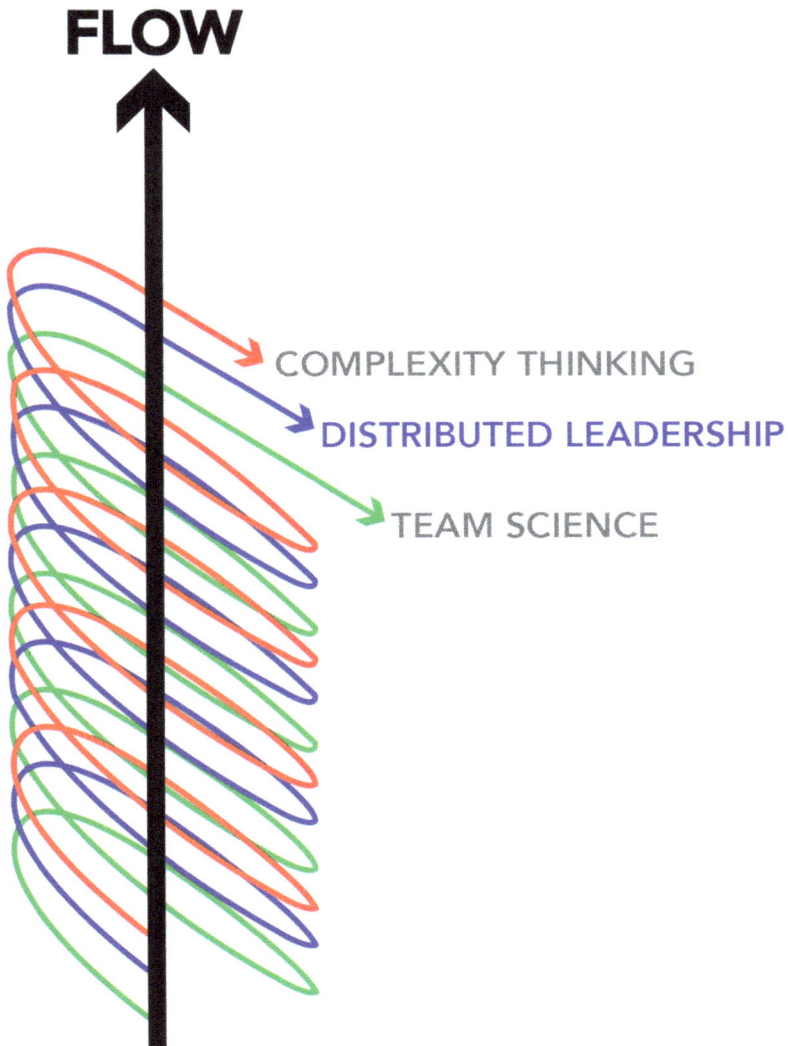

FLOW

COMPLEXITY THINKING

DISTRIBUTED LEADERSHIP

TEAM SCIENCE

CHAPTER 7

A Word About Leadership

L eadership matters (Turner and Baker, 2018). And it matters greatly in today's global landscape (Dorfman et al., 2012), and it influences countries, organizations, and teams (Antonakis and House, 2014). Leadership is essential in motivating followers, in providing meaningful and purposeful work, and in providing followers with the resources required to complete their jobs. Leadership has been attributed to affecting organizational innovation, adaptation, performance (Antonakis and House, 2014), and strategy. Although the discipline of leadership is one of the largest and most utilized areas of research, it still has room to grow.

Many critics of the field of leadership have noted its failure to produce effective and ethical leaders. For example, before the 2007 financial crisis in the United States, leadership was focused on short-term gains while instilling a culture of not challenging the status quo or questioning strategy and current practices (MacKenzie et al., 2014; Turner et al., 2018a). Current leadership practice has been viewed as holding onto the past, keeping with Tayloristic views of leadership that originally were established in the Age of Steel (Kersten, 2018). This can be problematic as organizations remain "siloed from the business, functionally specialized, and disconnected from each other" (Kersten, 2018: 5) rather than being distributed and flexible in response to adapting to environmental perturbations. These points are highlighted in the following description contrasting leadership and software–technology development:

> Business is further losing the ability to see or manage the work that the technologists are doing. Leadership seems to be using managerial tools and frameworks from one or two technological ages ago, while the technologists are feeling the pressure to produce software at a rate and feedback cycle that can never be met with these antiquated approaches. The gap between the business and technologists is widening through transformation initiatives that we're supposed to narrow it. We need to find a better way. (Kersten, 2018: 26)

This gap has extended to other industries beyond just technology and software (e.g., research and development, innovation, rapid product development). To highlight the problems with today's leadership practices in an innovative setting, as one example, Figure 7.1 shows the impediments to achieving each team's goals. At the time, these teams were developing various services and products to meet the customer's needs of a multinational financial corporation. These impediments were aggregated from several different groups involving nearly 1,000 employees, including five separate sessions from upper management. The impediments from these product development sessions were synthesized into two different categories: leadership and technology. The nature of the technology impediments (shown in Figure 7.1) were expected given that the nature of the team's work was software and technology based. The leadership impediments (Figures 7.1 and 7.2) outweighed any other impediments experienced by the team members. This ratio was unofficially 70/30%, with most of the impediments coming directly from leadership and the remaining coming from technology issues.

These results are repeatable in that similar findings have been experienced by most of the team development sessions that we have conducted for this same customer. Most teams listed leadership as being their number-one impediment, followed by categories identified as general, technology, and customer. Examples of the impediments for leadership include commitment from leadership, lack of leadership engagement, lack of timely communication from leadership, lack of clear vision, empowerment, and resistance to embrace change. Examples of general impediments include lack of resources, interruptions, team morale, and too many meetings. The technology and customer impediments are more specific to the organization's work and cannot be shared at this time. The main point is that leadership produces the most impediments from each product development session, time and time again. As we continue to engage, it has been our experience that teams from other companies and industries identify the same problems and impediments, with leadership leading the list.

An even more serious issue found with leadership is that they too often can create an environment that can be classified as unfriendly, resulting in a *culture of fear*. For example, when the results from the previous example (Figures 7.1 and 7.2) were shown to a member of the leadership team, this leader viewed the impediment listed on the chart and, referring to one of the sticky notes, asked: "Who wrote this?" As highlighted in Chapter 4, Systems and Complexity Thinking, leaders tend to blame others when confronted or questioned. This behavior produces a perception by the followers that they cannot freely speak their mind and question leadership. Providing a psychologically safe environment and connecting followers to their purpose (Quinn and Thakor, 2018) are essential elements of leadership. Leadership needs to empower their followers while also being engaged in their followers' work, providing them with the resources needed, and supplying them with an endless source of focus and direction. Unfortunately, as shown in the previous examples, reality and conceptualizations of leadership are often in a state of misalignment.

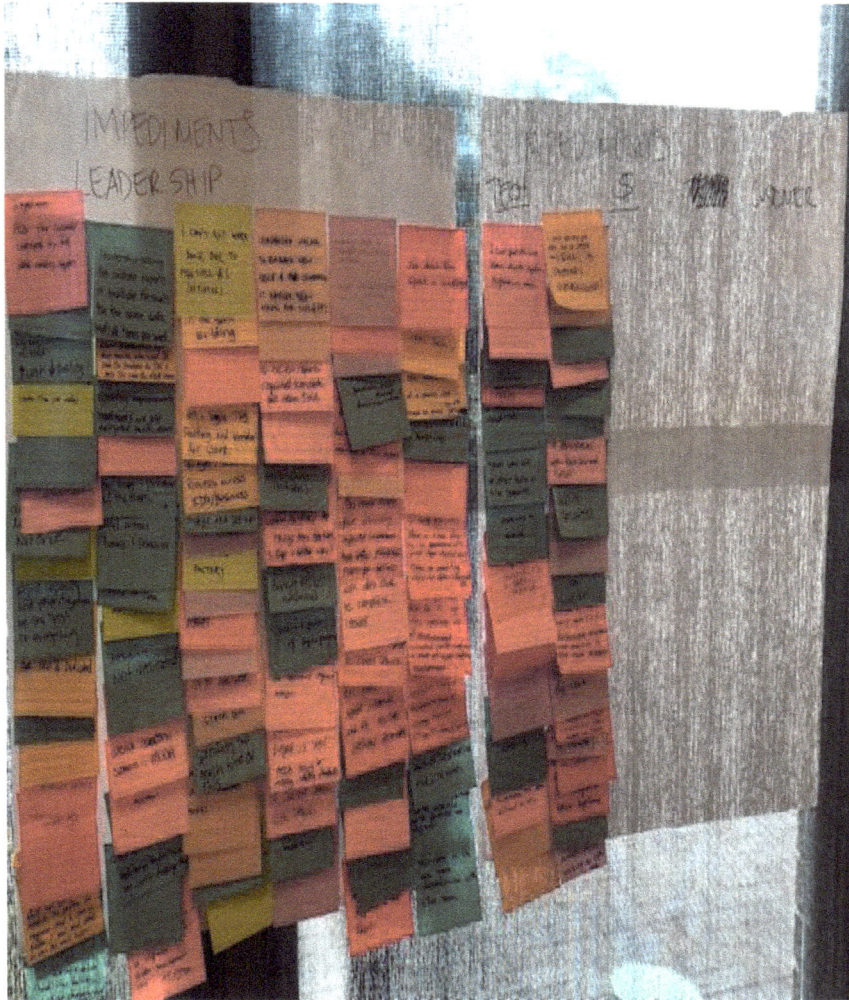

FIGURE 7.1. Leadership Impediments (left-hand chart) and Technology Impediments (right-hand chart).

Leadership identity must be in place at all levels. A leader must self-identify as a leader, as much as a leader's followers must identify them as a leader. When a leader does not self-identify as a leader, or when followers do not identify with their assigned leader, leader identity is in a state of resistance. Leader identity from both perspectives, the leader and the followers, is essential; a leader must develop his or her own identity as a leader, and followers must identify their assigned or appointed leader as a leader for them. This point was made clear in the product development sessions during which managers identified their

FIGURE 7.2.   Leadership Impediments

impediments, many of which related to leadership. These managers self-identified as leaders but still identified key impediments as being leadership impediments, not realizing that they were the very leaders causing the impediments that they had identified. This shortcoming in leader identity is one example showing how leadership, in today's organizations, needs to evolve from its current condition.

What is expected of leadership is often counter to what is practiced. This could be a factor of work overload, time constraints, operating between geographically dispersed locations, viewing managers as leaders, and a number of other factors that contribute to leadership trying to survive in complex environments.

Leadership theories have been identified by some as being composed of "concepts that are ill-defined, tautological, ideological and resist rigorous study" (Alvesson and Kinola, 2019: 12). Leadership theories are in need of more pragmatic solutions that produce more than just "upbeat ideologies fueling fantasies of the morally grounded, ethical, good, powerful leader" (Alvesson and Kinola, 2019: 1) as the central focal point. The field of leadership has been called to expand beyond just addressing organizational issues, it also must be responsible for implementing solutions to complex social issues (Antonakis and House, 2014). Deficits have been highlighted in leadership theories, showing a lack of the following:

- strategic structuring and planning,
- providing direction and resources,
- monitoring external environment, and
- monitoring performance and feedback (Antonakis and House, 2014).

Other calls have focused on extending leadership from the individual to more collective models of leadership (House, 1999), with features that account for leadership as a multilevel (Dionne et al., 2014; Hiller et al., 2011), multidimensional (Nielsen et al., 2016), and networked (White et al., 2016) construct. Others have called for combining leadership theories and to incorporate indigenous and global perspectives (Turner et al., 2018c) to appeal to those beyond the traditional Euro-American regions (include emerging countries concepts and needs).

At the bare minimum, leadership theories need to be able to address complexity (complexity thinking), must be multidimensional and multilevel (distributed leadership), and must be collective and networked (team science). Leadership needs to provide pragmatic theories and tools to operate in tomorrow's complex environments. These components make up The Flow System (complexity thinking, distributed leadership, team science). Leadership in complex environments cannot come from one individual; leadership must practice complexity thinking by using distributed leadership models or theories and by utilizing the benefits of team-based systems. Operating one of the three concepts without the other two will not produce flow, that is, an effortless in-the-moment process delivering value to the customer. All three must be intertwined as the triple-helix model indicates.

Distributed leadership provides a new perspective in that leadership is a team or group level construct with the team or group members (the individuals) as leaders. This point is essential to distributed leadership and to The Flow System. This chapter begins by making the distinction between leaders and managers, followed by differentiating between leader and leadership and the different levels of analysis (individual and team). A discussion on leadership development follows with a brief example of a team-based leadership development model.

Leadership capacities are introduced followed by a discussion about leadership in complexity. Last, we highlight current trends in leadership theory and identify the need for combining existing leadership theories to produce more practical and pragmatic solutions for complexity.

Leadership Versus Manager

Leadership and management, or leader and manager, often are used interchangeably (Toor and Ofori, 2008), but there are distinct differences between the two. Leadership has been conceptualized in multiple ways:

- as a dynamic and social process, an ability or skill, and as the relationship between leader and follower (Edwards and Turnbull, 2013),
- as being complex and multifaceted (Shuck and Herd, 2012), and
- as "a process whereby an individual influences a group of individuals to achieve a common goal" (Northouse, 2019: 5).

In contrast, managers are more structured in that they utilize specific tools and techniques to solve organizational problems and issues (Toor and Ofori, 2008) and to reduce chaos in organizations (Northouse, 2019). Management has been associated with a bag of tricks condensed into a few essential principles:

1. Management is about human beings. Its task is to make people capable of joint performance, to make their strengths effective and their weaknesses irrelevant. This is what organization is all about, and it is the reason that management is the critical, determining factor.
2. Because management deals with the integration in a common venture, it is deeply embedded in culture.
3. Every enterprise requires simple, clear, and unifying objectives. Its mission has to be clear enough and big enough to provide a common vision. The goals that embody it have to be clear, public, and often reaffirmed.
4. It is also management's job to enable the enterprise and each of its members to grow and develop as needs and opportunities change.
5. Every enterprise is composed of people with different skills and knowledge doing many different kinds of work. For that reason, it must be built on communication and on individual responsibility.
6. Neither the quantity of output nor the bottom line is by itself an adequate measure of the performance of management and enterprise. Market standing, innovation, productivity, development of people, quality, and financial results—all are crucial to a company's performance and indeed to its survival. . . . Performance has to be built into the enterprise and its management; it has to be measured—or at least judged—and it has to be continuously improved.
7. Finally, the single most important thing to remember about enterprise is that there are no results inside its walls. The results of a business is a satisfied customer. (Drucker, 2006: 194–196)

Management is associated with order and stability, whereas leadership is associated with adaptive and constructive change (Northouse, 2019). We highlight the differences between the two because they have distinct differences, especially when addressing complexity.

Operating in complex environments and addressing complex problems requires leadership. This leadership, in most cases, will require new models as existing leadership practices are too similar to the practices of management. In most organizations, it is difficult to distinguish between leaders and managers. Although some leaders can be managers, it is rare that true managers are leaders: they play different roles within an organization, and the two rarely merge into one. New leadership is required for complex environments, but the majority of resistance during this transformation will come from management.

Remember, managers like order and stability. A transformation to a new style of leadership and operating in complex environments naturally will result in resistance from current management. This resistance needs to be expected and should be built into any organizations' transformation. Understanding where this resistance comes from, and why, provides a good starting place for any organization.

This information highlights the general roles for managers, and the following sections discuss leaders and leadership further, followed by more detailed information about what new leadership might look like for this transformation.

Leader Versus Leadership

Looking at the terms leader and leadership, leader suggests one individual, "a person or thing that leads" (dictionary.com), whereas leadership represents the "act of leading or having the ability to lead" (dictionary.com). This distinction between the person or thing that leads and the act of leading is an important distinction. Leadership theories have been reviewed and identified as having two main elements in common: the locus and mechanism of leadership (Eberly et al., 2013) (Hernandez et al., 2011). The locus of leadership is best associated with the "source from which leadership originates; this can be the leader, the followers, the leader-follower dyad,[1] the larger collective (e.g., group of individuals, and entire organization), and/or the context" (Eberly et al., 2013: 427). The mechanism of leadership describes how leadership is being practiced. Mechanisms of leadership can come from direct means (e.g., direct leadership behavior, direct

---

1   Although a team is defined as consisting of two or more individuals, some research makes the distinction between an individual (one person), a dyad (two-individuals), and a team (more than two individuals). Many leadership theories also make this distinction (e.g., leader-member exchange, LMX; situational leadership). We define leadership as being two or more individuals (see Chapter 10, Teams, Teamwork, and Taskwork).

communication) or through indirect practices (e.g., affective, cognitive; Eberly et al., 2013).

The majority of leadership research has focused mostly on the direct or formal designated leader, indicating one in a position of leadership because of their hierarchical ranking within an organization. This unfortunate shortcoming in leadership research has resulted in a correction, and research is now beginning to expand to include more collective, nonhierarchical, types of leadership: "Leadership is no longer regarded as strictly hierarchical and nested within one person" (Eberly et al., 2013: 429). Current views of leadership are moving away from top-down, hierarchical, models to shared and collective forms of leadership. These new forms of leadership are aligned more closely with today's leadership, producing better models and theories of leadership for the real world: "Theorists have begun to conceptualize leadership as a broader, mutual influence process independent of any formal role or hierarchical structure and diffused among the members of any given social system" (DeRue and Ashford, 2010: 627).

These distributed forms of leadership are essential when operating in complex environments. No one individual is able to lead in complexity because of the unknown-unknowns, the ambiguity, the ill-defined problems, and the inability to predict with any certainty. Complexity required collectives, not individuals, to lead. From this perspective, the collective represents leadership, composed of individuals sharing in their role as a leader. Through the concept of distributed leadership, leadership is a shared property of the team or small group in which all team members participate in the leadership process (DeRue and Ashford, 2010).

Leadership as a collective, for example as a team, is successful only when the individual leaders, the team members, share information and communicate together so that they have a shared vision, a shared goal, and a level of shared cognition that is consistent and accurate among all members. This type of leadership, at the collective, requires effective teamwork to be present to achieve this level of sharedness that is required. Teamwork will be discussed further in Part IV, Team Science.

When viewing leadership remember that leadership occurs at multiple levels within an organization (individual, team, department, organization, and society). Leadership must extend horizontally, vertically, and everywhere in between, resulting in multiple agents playing an essential role in the overall leadership effort. This new multilevel, multiagent, and multidimensional role of leadership is what we call distributed leadership. Leadership is influenced by, and influences, every agent at all levels. This distributed, or networked, type of leadership is essential when addressing complexity. To be transformational, leadership development to achieve this distributed leadership model also will be required.

Leadership Development Versus Leader Development

Current leader development programs include more formal education designed for individuals, are too linear, and do not include a focus on teamwork or distributed relational interactions (agent interactions; Bolden and Gosling, 2006). Leader development involves individuals who are learning different leadership theories that may, or may not, be applicable to their workplace. The theories learned are designed to provide a general overview of different types of leadership models so that the individual is better able to alter his or her leadership style depending on the context and situation faced in the real world. Leader development does not provide tools and techniques that are designed specifically to an individual's place of work.

Other problems with past leader development programs are that little to no research has been conducted to assess leaders' maturity and performance throughout their careers (Muir, 2014). This has resulted in a lack of information about whether these programs are effective at developing leaders.

Leadership development is needed now more than ever as opposed to leader development. Leadership development programs need to provide tools and techniques that are relevant to the contextual conditions that leaders will face in their place of work along with addressing teamwork skills (the collective aspect of leadership).

Too often, organizations hire a leader without considering the contextual environment in which they will be operating, or they implement leadership development programs off-the shelf (i.e., those programs that have been designed for no one). These *cookbook approaches* to leadership development typically offer training using one intervention followed by a sequential set of additional interventions, regardless of their importance to the context or environment (Hanson, 2013). Leadership development efforts need to provide the tools and knowledge for future leaders to adapt to environmental variation that address today's complex environment, that focuses on collectives, that meet the needs of innovation, and one that provides a safe environment for leaders and employees to ask questions about strategy and procedures (Turner et al., 2018c).

One primary difference between leader and leadership development is that leadership development is collaborative whereas leader development focuses on the individual (Turner et al., 2018c). Leadership development results in a collective activity involving the capacity for one to "collaboratively work with others and focus on shared networks and meaning" (Turner et al., 2018c: 540; see also Gagnon et al., 2012). Leadership development efforts are essential for any organization attempting to transform into a more adaptive organization in this age of globalization and complexity. Leadership development crosses the divided question of whether leaders are born or made: "Leaders are not born, nor are they made; instead, their inherent potentials are shaped by experiences enabling them to develop the capabilities needed to solve significant social problems" (Mumford et al., 2000: 24). The next section highlights one example of utilizing team-based systems for leadership development: the Team Emergence Leadership Development and Evaluation (TELDE) model (Turner and Baker, 2017).

The Team Emergence Leadership Development and Evaluation Model

power class

Leadership development is best accomplished when the process of leadership is being practiced. Toyota's culture is one of development as opposed to training—development requires practice and is considered on-the-job development (OJD).[2] Training is too shortsighted whereas development is more closely aligned with Toyota's culture of continuous improvement, especially when it comes to the development of its employees. Essential practices to leadership development have been identified as follows: "(1) job assignments that provide exposure to novel, challenging problems; (2) mentoring; (3) appropriate training; and (4) hands-on experience in solving related problems" (Mumford et al., 2000: 24).

One technique in delivering each of these practices for developing leaders is to use a team-based system, such as the TELDE model (Turner and Baker, 2017).

The TELDE model is introduced in Figure 7.3; it provides a new way of viewing leadership development by "incorporating naturally occurring team processes as a means of replicating the characteristics traditionally viewed as being related to leadership development" (Turner and Baker, 2017: 3).

Leadership is emergent and considered an outcome that results from the interactions among individual members (Taylor et al., 2011). With leadership being the outcome, the team becomes the model of leadership, with individual team members as leaders. The whole, the team as the model of leadership, is an example of looking at leadership as a complex adaptive system (complexity thinking).

The TELDE model presented in Figure 7.3 is broken down into a team with four team members (left-hand side, vertical), which are represented as Team Member #1, Leadership Role, to Team Member #4, Leadership Role. This example uses a team task that also has been broken down into four episodes (bottom, horizontal), which are represented as Task Episode #1 to Task Episode #4.

The idea is to assign one team member as the leader for one task episode. This leadership role will be assigned to a team member that has experience and knowledge of how to complete the assigned task episode. The assignment for the next task episode will be given to a team member who has the skills and knowledge to complete that task episode, and so on until each team member has taken a leadership role or the task has been completed. No team member should take a leadership role twice until all other team members have taken a leadership role at

---

2   On-the-job development (OJD), on-the-job training (OJT), or training within industry (TWI) are similar terms used in the literature, sometimes synonymously. As with Toyota, TFS sees this as leadership development, not leadership training.

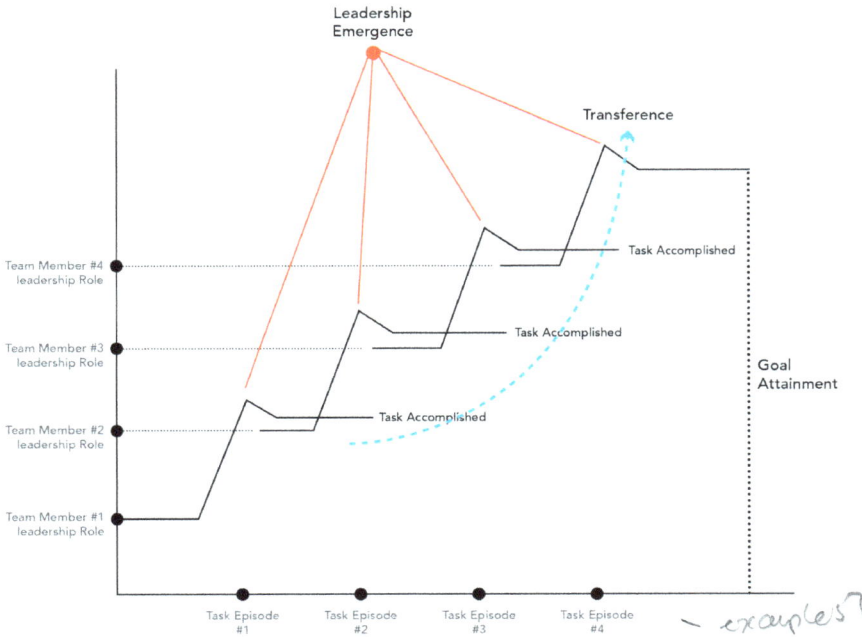

FIGURE 7.3. The Team Emergence Leadership Development and Evaluation (TELDE) Model

least once. If a new team member is inexperienced, then a mentor can be assigned to aid this team member through their role as a leader. This would be common for new employees.

The assignment of taking on a leadership role is an important stage in the leadership development process in that it requires developing a team's transactive memory system. Transactive memory systems originated from Wegner's research on married couples (Wegner, 1986), which later extended to team and small group research, described as team members knowing who has which skills, knowledge, and experiences within their team (Liang et al., 1995).

At the beginning of any team formation, it is essential for team members to become familiar with the other team members, knowing who has which skills, knowledge, and experiences as well as in which department each team member resides (if cross-functional team) and who has which level of ranking. This information is critical for team members to know so that each member knows who to go to when a specific knowledge or skillset is required to complete a task. In this example, using the TELDE model, knowing each other's knowledge, skills, and experiences aids in deciding who will take the leadership role for which task episode.

As each team member takes on a leadership role, consider Team Member #1, Leadership Role, shown on the left-hand side of Figure 7.3. This team member will be the leader for Task Episode #1, shown at the bottom of Figure 7.3. As this team member practices being a leader when addressing a real-world problem, they begin to grow as a leader. This growth is indicated by the upward movement in the line in Figure 7.3. This first line moves horizontally first to indicate a leader practicing sensemaking in trying to understand the task and how to direct the other team members in achieving this task. Once this team member understands the problem, they will begin planning, communicating, and assigning subtasks to other members, this is where the leader's growth in learning by practice is shown in the upward movement in the curve for Team Member #1. A level of confidence occurs in which the team member gains some level of identity as a leader. Once the task episode has been completed, there is a period of reflection, a debriefing mode, in which team members discuss items, such as what worked and what did not work. The growth as a leader for this team member may drop slightly while they reflect on how to become a better leader the next time they are called to lead. This drop is indicated in the downward slope for Team Member #1 shown in Figure 7.3.

Once the task episode has been completed, the next team member who is assigned as leader for Task Episode #2 follows. During each phase, the team members who are not leading also learn from the experience as they observe the team leader and make decisions about what they did that was effective and what was ineffective. Team members will learn from observing their peers during their leadership role and from participating in the reflection and debriefing stages of the process.

Ultimately, as each team member grows as a leader, the team achieves results in the model of leadership in which the task has been completed and in which leadership development took place (see *Transference* in Figure 7.3). This TELDE model utilizes each of the stages highlighted by (Mumford et al., 2000); real-world problems were addressed that were challenging and possibly novel, mentoring took place through the reflection and debriefing stages, and leadership development training was experienced by each team member. This example showcases just one method of leadership development that utilizes components of complexity thinking, distributed leadership, and team science. In the next section, we look at leadership capacities, examine leadership in complexity, and review current trends in leadership theory.

Leadership Capacities

In trying to uncover the mysteries of leadership, researchers have tried to answer such questions as: *What makes great leaders?* and *What makes great leaders effective?* Results have varied with some degree of success: "the quest has only been moderately successful" (van Knippenberg and Hogg, 2003: 244). Leadership research has identified specific traits, characteristics, behaviors, and practices that have been attributed to one being a successful leader within a specific contextual setting. Leadership theories also have been developed to explain how leadership functions as a process. In total, with generations of knowledge obtained from the field of leadership, the same questions remain with only a partial understanding of what constitutes a great leader.

Research has been conducted, in a number of different disciplines, that has highlighted the main traits (genetic such as height), characteristics (features such as personality), behaviors (actions taken), and practices (routine behaviors) of leaders. When we talk about capacities in this chapter, we are referring to the aggregate of these traits, characteristics, behaviors, and practices. One example includes research that identified six key capacities of effective leaders: "(a) More Global (b) More Innovative, (c) More Strategic Thinking (d) Better communicators, (e) More Effective Mentors and Developers of their People and (f) More Reflective and Proactive" (Watkins et al., 2011: 218). Others have identified capacities to include building dialogue, relationships, and networks (Grandy and Holton, 2013); instilling an appreciation of work; developing trust and cooperation; being flexible and establishing a meaningful identify (Weinberger, 2009); and engaging in difficult work (Ligon et al., 2011).

Capacities change as a leader develops and advances to higher levels within an organization or institution (Mumford et al., 2000). This means that different capacities are required to be developed at various stages of each individuals' career. Identifying the capacities required for each and every stage in one's career is nearly impossible. Leadership capacities from the literature have been reduced to 31 key categories. Of these 31 categories, specific leadership capacities can be selected based on the needs of the team or organization and based on the stage of development that the team or leader may be at. These 31 categories along with the capacities found in the literature are provided in Table 7.1. Table 7.2 provides a list of negative leadership capacities: these are the capacities that leaders should avoid. Known as the 32nd domain, these negative capacities identify toxic leaders that have resulted in "corruption, hypocrisy, sabotage, manipulation and other unethical behaviors: (Turner et al., 2018c; see also Edwards et al., 2015). For the purpose of leadership development, it becomes critical to first identify which leadership capacities are required for your specific organizational need. Then, and only then, should a leadership development program or tool be selected that is capable of developing the selected leadership capacities. This is counter to most leadership development practices that push development for preselected capacities that are not designed for your organization's specific needs.

TABLE 7.1. Leadership Capacity Categories

Leadership Categories	Leadership Capacities
Critical-Thinking Skills	Analytic skills, conceptual skills, convergent thinking skills, critical-thinking skills, divergent thinking skills, high level of technical skill and knowledge, knowledge,[a] knowledge-objective, meaning structures, mental models, practice clear thinking, pragmatic, prioritizing, problem or opportunity definition skills, questions assumptions, reasoning skills, scanning and analysis skills, screening, sorting, synthesize complex data, thinking systemically
Change	Change, lead change,[a] make change happen, change management, more change oriented, identify opportunities for change, leading change, understands importance of change processes
Coaching/Mentoring	Accept guidance, coach,[a] coaching,[a] effective teacher, invite coaching, invite guidance, mentor, mentor roles, mentoring, mentoring abilities, more effective mentors
Community Focused	Community development, develop community-inspired goals
Competencies	Ability to learn, business acumen,[a] cognitive ability, competent, corporate culture awareness, delivery skills, domain expertise, expert, expertise-objective, expertise-subjective, extraordinary capabilities, field knowledge, financial acumen, fundamental competencies, identify competency strengths, identify competency weaknesses, individual and organizational capacity, intellect, learn, organizational knowledge, planning skills, talented, technical proficiency;[a] technical skills
Conflict	Avoidance of conflicting relationships, conflict management,[a] conflict negotiation, conflict resolution, provide assistance with problems and conflict, resolve conflict[a]
Culture/Diversity/ Identity	Aware of cultural differences, awareness of workplace violence, cultural sensitivity, create a culture of support, cross-cultural knowledge, diversity, establishes meaningful identity, identity negotiation, leveraging diversity, member diversity, sense of social justice
Customer Service Oriented	Customer-centric strategy, customer focus, customer focused, customer service, customer service skills
Decision-Making Skills	Decision-making capacities,[a] decision making capabilities, encourages flexibility decisions and change, follow decision protocol, involve employees in decision making, make informed decisions
Developmental Skills	Creates visibility and momentum to move careers forward, develop a creative workforce, develop creative environment, develop mind-set for continuous learning, developing, developing talent, encourage employees' growth and development, growing skills, have development skills, identify opportunities, organizational and personal transformation, potency development, self-development, support career development
Emotional Intelligence	Be sensitive to needs of others, compassion, considerate of individuals, emotion, emotional,[a] emotional awareness,[a] emotional intelligence,[a] express emotions, know him/herself, self-awareness,[a] self-development, self-knowledge, self-verification
Entrepreneurial	Entrepreneurial learning, entrepreneurial spirit[a]

Feedback	Develop spirit of accountability, ensure accountability, evaluate individual contributions, feedback,[a] feedback seeking, multisource feedback, organizational feedback, provides evaluation and feedback, provides feedback, receptive to organizational feedback, seeks feedback from subordinates
Global Orientation	Environmental influence, global orientation, global perspective, global thinking, more global, shaping the environment
Innovative/Creative	Creative,[a] creativity,[a] innovation,[a] innovative,[a] innovativeness, innovator, more innovative, understand opportunities/implications of technical innovation
Leadership Qualities	Accept guidance, allocate adequate power, articulate direction, avoid blaming for shortcomings or failures,[a] belief system, challenge seekers, challenging, clarity of vision, complexity, create safe work environment, create vision, desire to lead, envisions future states, fair, follow, guides followers, idealized attributes, idealized behavior, idealized influence, identify breakthrough opportunities, identity as a leader, implementing a vision, innovative role modeling, instills knowledge and appreciation of work, know when to lead and follow, know when to be direct and collaborative, lead by example,[a] leader skills and knowledge, leader supportive, leadership perceptions, leading people,[a] learner autonomy, manage impressions, no public criticism, power, risk-taking,[a] role model, shape understanding of others, shared leadership, shared vision, supportive climate, uncertainty, vision[a]
Listening and Communication	Better communicators, common language, communicate,[a] communicate clear message, communicates expectations, communicating goals, communicating through open and honest dialogue, communication,[a] communication is encouraged, communication skills, creating open communication, depth of communication, fluency, listen, listen and communicate effectively, listens effectively, open, open communication,[a] openness to multiple perspectives,[a] whole listening
Management Functions/ Skills	Achieving results, act as buffer, affective commitment, align direction, align expectations,[a] broker, build organizational capacity, business performance, clarify vision, commitment to goals, continuance commitment, coordinator, creative process skills, delegates tasks, deliver results, detect opportunities, direct day-to-day activities, director, drive organization into future, driving execution, engage senior leaders, ensure role clarity, establish direction, excellence, expectation alignment,[a] facilitator, focus on quality, focus on results, foster adaptive behavior, foster co-evolution, functional, generates cooperation, goal directed, goal setting, identify emerging trends, identify needs, identify organizational champions, improve economic performance, impulse control, job instruction training, long-term focus, make task connections, makes resources available, manage, manage in uncertain situations,[a] manage first impressions, manage internal and external relations,[a] manage requests and constraints, manage resources, manage work, management by objective, managerial skills, managing talent, map workforce, monitor, negotiation,[a] normative commitment, optimizing fit,[a] organizational demand, organizational support, planning, project management skills, predict, prevent crises, producer, program management, promoting persistence, provide clear reporting structure, provide levels of autonomy, provides clear expectations, provides resources, purposeful action, results driven, secure organizational resources, seeking commitment, sell ideas, sense of urgency, set high standards, stakeholder success, stress management, sustained support, total quality management,[a] treat employees fairly, uncover needs of others, work-life balance, works cross-functionally

(Continued)

TABLE 7.1. Leadership Capacity Categories **(Continued)**

Leadership Categories	Leadership Capacities
Moral/Ethical	Ethical,[a] moral,[a] moral courage, values
Motivational	Congratulatory, empower peers to develop abilities, empower subordinates, encourages adaptation, encourages followers to try new approaches, encourages innovation, encourages self-regulation, energizing, engage desire for personal development, engagement, engaging and involving others, engaging and inspiring, excitement an motivational, generates confidence, generates enthusiasm, generates excitement, influential,[a] influence customers, influence decision makers, influence gatekeepers, inspirational,[a] inspiring commitment, instrumental behavior, intellectual stimulating,[a] motivational,[a] organizational rewards, recognition,[a] reinforce successes, reward,[a] stimulating, supportive,[a] supportive behavior, supportive work environment, uses intrinsic and extrinsic motivators
Networking	Dyadic interaction, encourage networking, framing of interaction patterns, network conditions, networking, modes of interaction
Organization Learning/ Learning Organization	Continuous learning,[a] organizational learning[a]
Performs Effectively	Performs effectively[a]
Individual Personality Traits or Characteristics	Adapt, adaptability,[a] adaptive,[a] admired, authenticity, aware, being in the moment, belief, commitment, compassion, confident,[a] considerate, consistent, courage,[a] courageous, curiosity, decisiveness, demonstrate personal energy, determination, determined, displays integrity, doer, driven,[a] dynamic, elaboration, enthusiastic,[a] extraversion,[a] fairness, flexibility,[a] flexible, general mood, honesty,[a] humility,[a] initiative, integrity,[a] intellectually gifted, intelligent, intuition, know strengths, know weaknesses, love of learning, obligation, optimistic, original, originality, overcome uncertainty, perseverance, persisent,[a] personal humility, personal quality, personality, persuasiveness, physically impressive, positive disposition, proactiveness, professional will, resilience, resourcefulness, respectful,[a] sees tasks and events as opportunities, self-disclosure, self-managed, self-confident,[a] self-direction, sound judgement, unconventional behavior,[a] unwavering
Political Acumen	Political acumen, political astuteness, political savvy
Problem-Solving Skills	Able to address ill-defined problems, creating the capacity to act, creative problem-solving skills, define executable problems, distributed view, idea evaluation skills, idea generation skills,[a] identify/plan/define problems, information processing skills, knowledge about complex problems, monitors deviations from standard, open to criticism, open to new ideas, outside-the-box thinking, problem-solving skills,[a] reframes problems, sensemaking, solicits solutions from followers
Reflective	Reflection,[a] more reflective and proactive, one-to-one reflection

Social/Relationship Building	Acceptance of individual efforts, awareness of personal values, belief in people, comradeship, develop skills in social relations, early relationship building, empathy,[a] focusing on the other, forging partnerships, foster collaboration, human resources management, human skills, importance of socializations, individualized consideration,[a] informal relationship building, interacts, interpersonal congruence, interpersonal skills,[a] make relationship connections, member involvement, more collaborative, more effective developers of people, one-on-one hierarchical relationship, participative behavior, partnering, partnership building, people management skills, people skills, practicing with others, relational awareness, relationship building,[a] relationship management,[a] sincere interest in others, sociable, social problem solving, social skills,[a] willing to let others take control
Strategic Thinking	Create strategic footing in organization, creating a strategic vision, develop strategy, high level of strategic-thinking capacities, more strategic thinking, provides a strategic vision, shaping strategy, strategic thinking, strategic vision
Teamwork/Team Building	Allow team members to make decisions, build teams,[a] build teamwork and consensus,[a] building coalitions,[a] building a shared vision, collaborative,[a] collective process, common mission, enhances team effectiveness, facilitate team interactions, foster collaboration, foster collaboration and teamwork, lead creative teams, learn about diversity of team members, maintains level of group harmony, managing team, orient teams toward goals, shared control, shared expertise, shared vision,[a] supports team, task interdependence, team learning processes, team playing, team spirit, transactive memory skills,[a] understands importance of teams
Trust/Trustworthiness	Commitment to the truth, create trust,[a] generates trust, honesty, trust,[a] trustworthiness,[a] trusted

Note: [a]Capacities listed more than once.

Source: Borrowed from Turner et al. (2018b, 2018c).

TABLE 7.2.   Negative Leadership Capacities

Negative Leadership Capacities

Attributional egotism, callous, control, corrupt, disregard for regulatory requirements, dominant culture, evil, excessive risk taking, gender power relations, gender-based obstacles, hypocrisy, incompetent, infects leadership, insular, intemperate, interrupts organizational learning, lack of feedback, lack of prior experience, manipulative, non-supportive work environment, opportunistic, poor working relationships, pursuing short-term objectives, rigid,[a] sabotage, self-deception, self-interest, sense of omnipotence, short-term focus, strong self-investment, thwarts performance, triumphant contempt, using processes where outcomes are unknown

---

Note: [a]Negative capacities listed more than once.

Source: Borrowed from Turner et al. (2018c).

Leadership in Complexity

In navigating complex environments, Richardson provided the following guidelines for both managers and leaders when practicing complexity thinking:

1. Just because it looks like a nail, it doesn't mean you need a hammer: A complex systems view acknowledges that context recognition is problematic, and as such deceasing what to do is not a simple exercise of repeating what you did the last time you were in the same situation. The chances are the situation is quite different.
2. Decisions made by the many are often better than those made by a few: A precursor to any decision has to be a thorough consideration (critique) from multiple perspectives (pluralism). This might be the application of a variety of different models, or simply just asking more than one person for their opinion. Such an approach quite naturally leads to creative thinking, and enables the development of a richer understanding on a context of interest before a decision is made. Beware, however, as "too many cooks may spoil the broth," and in situations where time is not readily available, the leadership of an individual may prove more effective than attempts at groups decision making;
3. Expect to be wrong (or at least not completely right): There are limits to how pluralistic and critical our decision making processes can be. But even with all the time and resources in the world (and a commitment to do the "right" thing), decisions can only be made based on our best current understanding, and that understanding will always be incomplete. Everything is connected to everything else. We can't consider everything so we construct artificial boundaries to help us make a decision-without those boundaries we are helpless, with them our responses are limited (but at least we have some responses!);
4. Flip-flopping is OK: Contrary to the beliefs of certain US politicians, being prepared and confident enough to change one's mind when it becomes clear that one's model is proving ineffective (and even counterproductive) is actually a virtue, not a sin. The complex organization evolves in unforeseeable ways and as such we must be prepared to "move with the times." The simple act of making a decision (based on past experience) can change how the future unfolds. Don't make the mistake of escalating one's commitment in the face of mounting contrary evidence. Dogmatism is rarely an effective strategy. (Richardson, 2008: 25)

Leadership Theory Trends

The field of leadership has evolved over the years along with most every other discipline (e.g., psychology, physics, sociology). During this evolution, leadership research and theories have been categorized into four different evolutionary life cycles, "ranging from individual trait and competency theories, to situational/contingency theories, to leader-follower (dyad) theories, to more collective theories including newer global and networking theories" (Turner and Baker, 2018: 480). These four categories of leadership theories have been identified as being traditional, newer, collective, and global leadership theory categories (Avolio et al., 2009; DeRue and Ashford, 2010). The following sections provide a brief description of each of these four categories.

Traditional Leadership Theories

Traditional leadership theories have been identified as those theories that dominated the literature up to the late 1970s (Avolio et al., 2009). Theories in this category include trait-based, behavior, leader-follower relationships, and contingency-type (reward and punishment) theories. Some of the more common theories from the traditional category include the great-man theory (which is a trait-based theory; Kirkpatrick and Locke, 1991), the managerial grid dimensions of production and concern for people (Hernandez et al., 2011), and situational leadership (Blanchard, 2010; House and Mitchell, 1974), as well as theories that began to look at leader–follower relationships.

Newer Leadership Theories

Newer leadership theories extended the research from the traditional category (late 1970s) to research from the late 1990s and into the early 2000s (Avolio et al., 2009). Theories in this category focus on symbolic behavior, vision, inspiration, charismatic, affective attributes, morals, individualization, and intellectual stimulation (Avolio et al., 2009). Some of the more common theories from this category include charismatic leadership, transactional and transformational leadership, and servant leadership.

Collective Leadership Theories

Collective leadership theories add an additional level of analysis to most traditional and newer leadership theories. Theories in this category view leadership from the collective perspective in which leadership is shared, distributed, and networked across agents. Leadership occurs in organizational structures that are less hierarchical and the roles of the leader change quite frequently: "One could be taking on leadership roles and responsibilities in one situation or at one time and then switch to a followership role in another situation or time" (Scott et al., 2017: 465). Collective theories include leadership as a multilevel construct, incorporate behaviors and skills beyond those included in traditional and newer leadership theories, and view leadership as a dynamic process that is socially constructed among agents.

Some of the common theories in this category include various team leadership models, such as the TELDE model previously discussed (Turner and Baker, 2017), shared and distributed leadership, implicit leadership theory (ILT; (Scott et al., 2017), complex leadership, self-leadership, and empowering leadership (Dionne et al., 2014).

Global Leadership Theories

Recently, researchers have begun looking at how leadership varies across geographic regions and how operating in multiple regions should be addressed by today's leaders. Global leadership theories contribute to the existing body of theories on leadership by focusing on "(a) the meaning of leadership as a cultural construct and (b) the variations in local expectations regarding leader behavior" (Steers et al., 2012: 481). Global leadership also extends to complexity, flow, and presence as highlighted in the following definition of global leadership: "The process of influencing others to adopt a shared vision through structures and methods that facilitate positive change while fostering individual and collective growth in a context characterized by significant levels of complexity, flow and presence" (Mendenhall et al., 2012: 500). Leadership theories that fall within this category include global and indigenous leadership theory's, instrumental leadership (Antonakis and House, 2014), and culturally endorsed leadership theory (CLT; Arvey et al., 2015).

New Directions in Leadership Theory

Beyond these four categories (traditional, newer, collective, global), other new directions for the field of leadership have been identified. Some of these new directions include addressing new contextual issues related to leadership, covering more ethical issues, leading for creativity and innovation, covering leadership in virtual settings, and better understanding destructive leadership behaviors (Gardner et al., 2010). These trends also have called for more theories of complexity leadership, leadership in teams and decision groups, and strategic leadership by top executives (Gardner et al., 2010). Leadership needs to be placed in the social system and viewed, not as an individual, heroic behavior but rather as a practice within a system that occurs at multiple levels (Pearce and Conger, 2003).

To achieve these new directions, many existing theories are being combined into newer *integrated* leadership theories. The field of leadership has come predominantly from European and U.S. institutions; some people have been calling for more theories to be developed that reflect developing and emerging countries. This creates a need for more global and indigenous leadership theories to be developed. As organizations around the globe are transforming to team-based systems, in one form or another, new leadership models need to be designed that can accommodate the growing needs resulting from so many of the problems organizations currently are facing as they attempt to manage such nested global structures. This need comes from theories within the collective category and will need to be combined with theories from other categories (traditional, newer, global) so that more comprehensive theories can be provided to organizations. Combining theories to meet new organizational and societal needs will be the trend moving forward.

This practice of combining theories, also termed integrated or *hybrid theories*, are recommended to advance the field of leadership (Meuser et al., 2016). More important, integrating existing theories into new hybrid models for The Flow System, to achieve *flow*, is needed to meet the needs of organizations and to provide unwavering and expedient value to the customer. The next two chapters will discuss potential leadership theories that could be integrated into a hybrid leadership model for such combined–blended hierarchical structures.

Conclusion

Leadership matters and achieving *flow* cannot be achieved without leadership. More important, distributed leadership is essential—that is, leadership that extends horizontally, vertically, and every place in between. Distributed leadership requires shared and team leadership models, while providing top-down and bottom-up processes that cross organizational boundaries and break down existing siloed structures. Barriers need to be removed so that leadership influences, and is influenced by, all agents and agencies as opposed to operating in strictly top-down, command-and-control, structures.

Today's organizations are being disrupted at alarming rates and complexity needs to be built into the leadership model along with leadership being a collective construct for today's team-based structures. Through self-realization, coaching, and insight many organizations have or are beginning to transition to team-based structures, utilizing tools from Agile and DevOps. Unfortunately, these efforts often fail because of a "lack of infrastructure and business buy-in to do so effectively" (Kersten, 2018: 26). One of the biggest impediments that we have encountered in the past, when trying to implement agile-type tools and practices, is that existing leadership hierarchical systems resist these efforts. This resistance, in part, comes from management at the executive levels. Managers do not want to release control to team-based structures. Unfortunately, to allow team-based systems to operate with autonomy and to be self-organizing, which is essential to become successful, management must release control over to the teams. A separation between management and leadership needs to take place.

As the team or collective becomes the model of leadership, this often threatens existing management agents, which becomes an impediment to any agile or *flow* initiative. Most agile-type initiatives fail because management prevents the team-based system from operating freely. They refuse to remove boundaries (complexity thinking), thus choking the system to the point at which it cannot function freely, managing an open system using closed-system tools. Leadership needs to be structured so that it allows the team-based system to operate freely, as an open system. Leadership still can maintain its hierarchical structure at the executive level, when desired, but just not at the team levels. This calls for more blended organizational structures using hybrid leadership models. Some of these hybrid leadership theories and models will be discussed in more detail in the following chapters.

References

Alvesson M and Kinola K. (2019) Warning for excessive positivity: Authentic leadership and other traps in leadership studies. *The Leadership Quarterly* 30: 383–395.

Antonakis J and House RJ. (2014) Instrumental leadership: Measurement and extension of transformational–transactional leadership theory. *The Leadership Quarterly* 25: 746–771.

Arvey R, Dhanaraj C, Javidan M, et al. (2015) Are there unique leadership models in Asia? Exploring uncharted territory. *The Leadership Quarterly* 26: 1–6.

Avolio BJ, Reichard RJ, Hannah ST, et al. (2009) A meta-analytic review of leadership impact research: Experimental and quasi-experimental studies. *The Leadership Quarterly* 20: 764–784.

Blanchard K. (2010) *Leading as a higher level: Blanchard on leadership and creating high performing organizations.* Upper Saddle River, NJ: FT Press.

Bolden R and Gosling J. (2006) Leadership competencies: time to change the tune. *Leadership* 2: 147–163.

DeRue SD and Ashford SJ. (2010) Who will lead and who will follow? A social process of leadership identity construction in organaztions. *Academy of Management Review* 35: 627–647.

Dionne SD, Gupta A, Sotak KL, et al. (2014) A 25-year perspective on levels of analysis in leadership research. *The Leadership Quarterly* 25: 6–35.

Dorfman P, Javidan M, Hanges P, et al. (2012) GLOBE: A twenty year journey into the intriguing world of culture and leadership. *Journal of World Business* 47: 504–518.

Drucker PF. (2006) *Classic Drucker.* Boston, MA: Harvard Business School Press.

Eberly MB, Johnson MD, Hernandez M, et al. (2013) An integrative process model of leadership: Examining loci, mechanisms, and event cycles. *American Psychologist* 68: 427–443.

Edwards G, Schedlitzki D, Ward J, et al. (2015) Exploring critical perspectives of toxic and bad leadership through film. *Advances in Developing Human Resources* 17: 363–375.

Edwards G and Turnbull S. (2013) A cultural approach to evaluating leadership development. *Advances in Developing Human Resources* 15: 46–60.

Gagnon S, Vough HC, and Nickerson R. (2012) Learning to lead, unscripted: developing affiliative leadership through improvisational theatre. *Human Resource Development Review* 11: 299–325.

Gardner WL, Lowe KB, Moss TW, et al. (2010) Scholarly leadership of the study of leadership: A review of *The Leadership Quarterly*'s second decade, 2000–2009. *The Leadership Quarterly* 21: 922–958.

Grandy G and Holton J. (2013) Evaluating leadership development needs in a health care setting through a partnership approach. *Advances in Developing Human Resources* 15: 31–82.

Hanson B. (2013) The leadership development interface: Aligning leaders and organizations toward more effective leadership learning. *Advances in Developing Human Resources* 15: 106–120.

Hernandez M, Eberly MB, Avolio BJ, et al. (2011) The loci and mechanisms of leadership: Exploring a more comprehensive view of leadership theory. *The Leadership Quarterly* 22: 1165–1185.

Hiller NJ, DeChurch LA, Murase T, et al. (2011) Searching for outcomes of leadership: A 25-year review. *Journal of Management* 37: 1137–1177.

House RJ. (1999) Weber and the neo-charismatic leadership paradigm: A response to beyer. *The Leadership Quarterly* 10: 307–330.

House RJ and Mitchell TR. (1974) Path-goal theory of leadership. *Journal of Contemporary Business* 3: 81–97.

Kersten M. (2018) *Project to product.* Portland, OR: IT Revolution.

Kirkpatrick SA and Locke EA. (1991) Leadership: Do traits matter? *Academy of Management Executive* 5: 48–60.

Liang DW, Moreland R, and Argote L. (1995) Group versus individual training and group performance: The mediating role of transactive memory. *Personality and Social Psychology Bulletin* 21: 384–393.

Ligon GS, Wallace JH, and Osburn HK. (2011) Experiential development and mentoring processes for leaders for innovation. *Advances in Developing Human Resources* 13: 297–317.

MacKenzie C, Garavan TN, and Carbery R. (2014) The global financial and economic crisis: did HRD play a role. *Advances in Developing Human Resources* 16: 346–356.

Mendenhall ME, Reiche BS, Bird A, et al. (2012) Defining the "global" in global leadership. *Journal of World Business* 47: 493–503.

Meuser JD, Gardner WL, Dinh JE, et al. (2016) A network analysis of leadership theory: The infancy of integration. *Journal of Management* 42: 1374–1403.

Muir D. (2014) Mentoring and leader identity development: A case study. *Human Resource Development Quarterly* 25: 349–379.

Mumford MD, Zaccaro SJ, Harding FD, et al. (2000) Leadership skills for a changin world: Solving complex social problems. *The Leadership Quarterly* 11: 11–35.

Nielsen MB, Skogstad A, Matthiesen SB, et al. (2016) The importance of a multidimensional and temporal design in research on leadership and workplace safety. *The Leadership Quarterly* 27: 142–155.

Northouse PG. (2019) *Leadership: Theory and Practice.* Thousand Oaks, CA: Sage.

Pearce CL and Conger JA. (2003) All those years ago: The historical underpinnings of shared leadership. In: Pearce CL and Conger JA (eds) *Shared leadership: Reframing the hows and whys of leadership.* Thousand Oaks, CA: Sage, 1–18.

Quinn RE and Thakor AV. (2018) Creating a purpose-driven organization. *Harvard Business Review* 96: 78–85.

Richardson KA. (2008) Managing complex organizations: Complexity thinking and the science and art of management. *Emergence: Complexity and Organization* 10: 13–26.

Scott CPR, Jiang H, Wildman JL, et al. (2017) The impact of implicit collective leadership theories on the emergence and effectiveness of leadership networks in teams. *Human Resource Management Review* 28: 464–481.

Shuck B and Herd AM. (2012) Employee engagement and leadership: Exploring the convergence of two frameworks and implications for leadership development in HRD. *Human Resource Development Review* 11: 156–181.

Steers RM, Sanchez-Runde C, and Nardon L. (2012) Leadership in a global context: New directions in research and theory development. *Journal of World Business* 47: 479–482.

Taylor A, Cocklin C, Brown R, et al. (2011) An investigation of champion-driven leadership processes. *The Leadership Quarterly* 22: 412–433.

Toor S-u-R and Ofori G. (2008) Leadership versus management: How they are different, and why. *Leadership and Management in Engineering* 8: 61–71.

Turner JR and Baker R. (2017) Team Emergence Leadership Development and Evaluation: A Theoretical Model Using Complexity Theory. *Journal of Information and Knowledge Management* 16: 17.

Turner JR and Baker R. (2018) A review of leadership theories: identifying a lack of growth in the HRD leadership domain. *European Journal of Training and Development* 42: 470–498.

Turner JR, Baker R, JSchroeder J, et al. (2018a) Leadership development techinques: Mapping leadership development techniques with leadership capacities using a typology of development. *European Journal of Training and Development* 42: 538–557.

Turner JR, Baker R, Schroeder J, et al. (2018b) Mapping leadership development techniques with leadership capacities using a typology of development. *European Journal of Training and Development* 42: 538–557.

Turner JR, Baker R, Schroeder J, et al. (2018c) The global leadership capacity wheel: Comparing HRD leadership literature with research from global and indigenous leadership. *European Journal of Training and Development* 43: 105–131.

van Knippenberg D and Hogg MA. (2003) A social identity model of leadership effectiveness in organizations. *Research in Organizational Behavior* 25: 243–295.

Watkins KE, Lyso IH, and deMarrais K. (2011) Evaluating executive leadership programs: A theory of change approach. *Advances in Developing Human Resources* 13: 208–239.

Wegner DM. (1986) Transactive memory: A contemporary analysis of the group mind. In: Mullen G and Goethals G (eds) *Theories of group behavior.* New York, NY: Springer-Verlag, 185–208.

Weinberger LA. (2009) Emotional intelligence, leadership style, and perceived leadership effectiveness. *Advances in Developing Human Resources* 11: 747–772.

White L, Currie G, and Lockett A. (2016) Pluralized leadership in complex organizations: Exploring the cross network effects between formal and informal leadership relations. *The Leadership Quarterly* 27: 280–297.

CHAPTER 8

Team and Distributed Leadership

O rganizational structures vary from one organization to the next. These structures also need to be altered when an organization moves from a traditional hierarchical structure to one that incorporates team-based structures. In many cases, when organizations utilize team-based structures, they do so without first altering the organizational structure under which the teams actually function. This leads to many failed attempts in implementing team-based structures, with claims that team-based systems or agile tools and practices do not work. The main point that management must acknowledge is that different reporting and leadership structures must be put in place when moving to team-based systems.

Blended and Hybrid Leadership Model

One example of a blended leadership model for team-based systems is shown in Figure 8.1. This figure highlights the different leadership styles required, with the top layer keeping the traditional executive leadership structure and the lower portion changing to the team-based structure. This blended, or hybrid, structure utilizes distributed and shared leadership models for the team-based systems, while also allowing organizations to maintain their existing leadership structure at the executive level.

The horizontal line in Figure 8.1 represents the blended structure line, the point at which the transition from the team-based system practicing shared leadership meets with the organization's existing hierarchical structure. The point at which this transition takes place is dictated by the organization and should be where the team-based structure ends. This point will vary from one organization to the next, but the blended structure provides organizations with a model that best serves team-based structures.

This blended structure, which is hierarchical at the executive levels and features team-based structures at the lower levels utilizing shared leadership, is what we are calling for. We are not calling to replace the original structure of an entire organization. This point has been highlighted by others in the field: "Shared leadership supplements but does not replace hierarchical leadership" (Pearce and Conger, 2003: 281).

In fact, not all divisions in an organization will require team-based systems. For example, divisions that operate within the simple domain and at the lower levels of the complicated domain will be best served using its existing structure. Only divisions or departments that encounter complexity, and in some cases, those operating at the higher levels of the complicated domain that run the risk of encountering complexity, should be restructured to utilize the benefits of a team-based structure using shared leadership. These simple points go back to complexity thinking and being able to identify the level of complexity that might be experienced and then planning accordingly.

The blended model that we are presenting here is similar to complexity leadership theory (CLT) (Uhl-Bien and Marion, 2009; Uhl-Bien et al., 2007), which provides a meso model to look at the interplay between an organization's formal administrative functions and its informal adaptive functions. This interplay is described as *entanglement*: "A dynamic relationship between the formal *top-down, administrative forces* (i.e., bureaucracy) and the informal, *complexly adaptive emergent forces* (i.e., CAS) of social systems" (Uhl-Bien et al., 2007: 305; emphasis added). Complexity leadership theory is an integrated theory that incorporates three organizational leadership systems, administrative, adaptive, and enabling, into one cohesive organizational meso model.[1] The difference with the blended model that we are presenting is that we are not suggesting a means to transform an entire organization, although one could transform an entire organization based on our blended model, if needed. The key concept is to implement our blended model where needed, for those areas within an organization that do deal with complexity while also utilizing the benefits of team-based systems. These two organizational models are similar but different. They do share, however, the same goals: to create emerging conditions "where innovation and adaptability are needed" (where complexity is being addressed) and to facilitate "the flow of knowledge and creativity" (Uhl-Bien et al., 2007: 305) from team-based structures.

This chapter, Chapter 8, Team and Distributed Leadership, will focus primarily on the lower portion of Figure 8.1. It looks at shared leadership and essential

---

1   In identifying different levels of a theory, four levels typically are identified, but not all four levels are utilized in any one theory. Some theories can entail just one level, whereas other theories could include all four. These levels include micro (individual or team), meso (departmental or organizational), macro (sector or industry), and meta (country or global). The divisions between each level are not absolute and can vary from one theory to the next. For complexity leadership theory, as in this example, the theory is concentrating on both micro and meso levels.

New Organizational Structure
(Required for Team-Based Systems)

Executive Level

BLENDED   **STRUCTURE**

Team-Based System

SHARED LEADERSHIP

Leadership

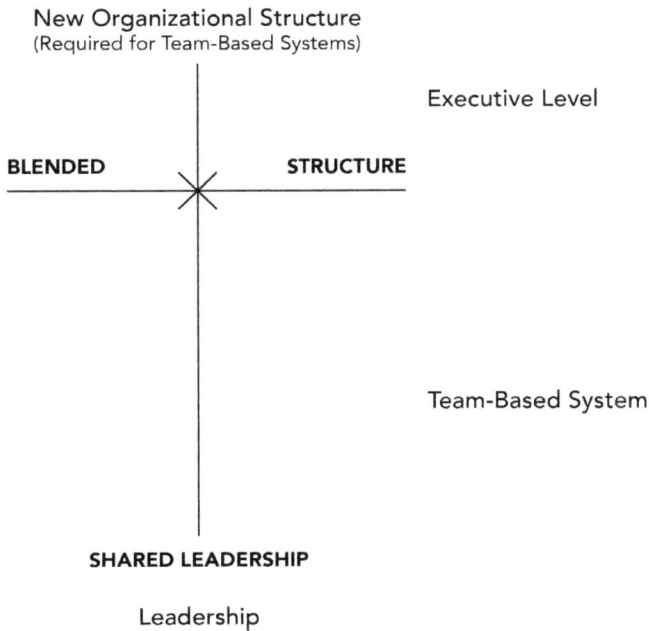

FIGURE 8.1.   Blended Leadership Model for Team-Based Systems

antecedents needed when implementing shared leadership. The next chapter, Chapter 9, Strategic, Instrumental, and Global Leadership, focuses on the top portion of Figure 8.1, the area above the blended structure line. Chapter 9 also introduces some leadership theories that have been found to work well with team-based systems. The aggregate, a blended structure utilizing shared leadership with components of strategic, instrumental, and global leadership, provides a leadership model for *flow* that we call distributed leadership.

Team Leadership

When viewing team leadership, one can take three positions.

- One could apply any standard leadership theory or model to a team.
- One could apply a leadership theory or model that is designed for a team.
- One could modify an existing leadership theory or model to address teamwork and the contextual setting, making it more applicable for a team setting.

Given that current research is just beginning to look into leadership at the team or collective level of analysis, there is a deficit in leadership theories that address teamwork and team effectiveness. This limits the choices for the second position. Applying current theories to a team (the first option) may work in many cases; however, when teamwork is essential, as in instances in which complexity is being experienced, teamwork needs to be included in the leadership framework. In most cases, to achieve desired results, implementing an existing leadership theory in a team setting will require some form of modification as highlighted in the third option. Because of the deficit in team leadership, theories that address teamwork and team effectiveness, beginning with the wealth of knowledge that comes from leadership research across multiple disciplines, is the best starting point (van Knippenberg, 2017).

When looking at team leadership, one question that is relevant to the conversation is whether or not a team needs a leader. In general, newly formed teams are in need of a strong leader, either internal or external, to get the team up and running. Once a team has experience working together and have a clear understanding of their goals and vision, leadership external of the team is necessary only when additional resources and information are needed for the team to accomplish its goal.

Team Leadership Variations

Team leadership could be conceived in a number of ways. Figure 8.2 provides a number of variations for conceiving team leadership. What is found quite frequently, when an organization first moves to a team-based structure, is that an external leader manages the team.

Figure 8.2(a) represents an external leader model that is closely associated with a directive style of leadership. Here, the external leader operates as the team's manager providing a top-down, or command-and-control, type of structure. The team reacts to the external team leader's directives. This type of structure eliminates any teamwork from occurring while also preventing creativity and emergence. Teams respond to direction rather than team members working interdependently and independently on shared tasks, which constitutes a team.[2] Top-down leadership of teams, as in this external team leader model, often results in micromanaging and is counterproductive to teamwork and team effectiveness. In most cases, it would be better to disband teams all together using this style of leadership. The advantages of using teams is counteracted by the top-down leadership, essentially making teams ineffective.

In contrast to the directive type of team leadership model, Figure 8.2(b) shows a recursive type of leadership structure. Here, an external leader is still overseeing the actions of the team, but this external leader also listens to the team members. This relationship, between the external team leader and the team members,

---

2   Teams and teamwork will be discussed in more detail in Part IV, Team Science.

is a recursive relationship rather than a command-and-control, top-down relationship, as shown in Figure 8.2(a). By being a recursive relationship, the external team leader not only provides direction and resources but also is responsive to team member needs and inquiries. The external leader provides an environment in which team members are able to question orders so that all members agree on their course of action to accomplish the team's goal.

This type of recursive relationship is similar to team coaching or participative leadership. In looking at leader assimilation processes, Manderscheid (Manderscheid, 2008) suggested external leaders need to solicit feedback. This feedback mechanism operates both ways, the leader selects a team member to solicit feedback from other team members, and this feedback is shared with the leader, which is followed by the leader responding to the team members in which a collective plan is set for moving forward. Some support from the literature suggests that team coaching results in better team performance and safer environments; however, this type of team leadership model does not have a clear framework (van Knippenberg, 2017). An additional caveat with this recursive type of team leadership model emerges when multiple teams are being led by one leader or coach. Having to manage multiple teams results in cognitive overload of the team coach, reducing the attention that the coach can give to any one team. This also places a strain on the resources available to the teams, especially when some resources need to be shared among the different teams. As more teams are added, an external leader is less capable of being responsive to team members' needs.

Figure 8.2 provides two models that are similar: Figure 8.2(c) presents an external team leader with an internal team leader, and Figure 8.2(e) identifies an internal dual-team leader model. In researching innovative teams, Folkestad and Gonzalez (Folkestad and Gonzalez, 2010) found that team leadership was necessary to monitor the teams progress and to change the makeup of individual team members, if determined necessary. In this capacity, Folkestad and Gonzalez (2010) found that a team leader's role consisted primarily of selecting team members, establishing goals, inspiring to achieve goals, evaluating progress, accepting responsibility, and maintaining the ability to make necessary changes. In addition to a team leader, Folkestad and Gonzalez (2010) identified the need for a second, high-profile team member who could make connections to groups external to the team. This division of roles places the team leader's role as one that concentrates on the inner working of the team with the high-profile team member's role being one that concentrates on external connections required for goal attainment. This can be achieved using either model presented in Figure 8.2(c) and 8.2(e).

For Figure 8.2(c), the team is led by an external leader and a team leader is assigned or appointed to account for external communications. For Figure 8.2(e), team members are assigned or appointed to both positions: the internal leader of the team and the external leader to oversee external communications. In both cases, team members are more involved in the activities and goals of the team and have more input in their roles.

Some drawbacks of the external team leader with an internal team leader, shown in Figure 8.2(c), are similar to those found in the recursive model in Figure 8.2(b). If an external leader is assigned to multiple teams, then the amount of time and resources available to each team is diminished, resulting in less than adequate coverage for each team.

Likewise, in Figure 8.2(e), as team members take on roles of being both the internal team leader and the external team leader, these roles take away from the team's tasks because two team members are working on team tasks as well as leadership tasks. This could result in lower performance for the team and also could result in some team conflict because of a few team members working on more tasks than other team members (the team leaders in this case). Whether true or not, this could be the perception of some team members. Having internal team members operating as external communicators with those outside of the team also could be problematic. This is likely to occur when a team member is required to communicate upward, with organizational members at higher levels in the organization. This hierarchical power differential, or power distance, could result in delayed responses from the organization, which could delay the team in achieving its goal. Coordinating one team's task with other teams could be problematic when multiple teams are working toward a common goal. This leadership model could be problematic in multiteam settings,[3] for example.

Figure 8.2(d) highlights a team that is led by one team member. This internal team leader can either be appointed, typically by management, or assigned by the team members. When a team leader is mentioned, this is the model that is most often conceptualized. One team member acts as the leader for the whole team. Although this team leadership model may work well for one team once they are familiar with the team member acting as the leader, this model is effective only when dealing with simple or complicated problems but is less effective when complexity enters the picture. A single leader, in this case the team leader, cannot "possess all of the necessary skills, knowledge, and abilities to lead all aspects of knowledge work" (Wang et al., 2017: 155), especially when the environment changes from complicated to complex.

Other issues may come up when a team concentrates its leadership functions internally, such as when a team is led by one team member, and coordination of activities external to the team becomes harder to manage. With no external leader ensuring that the team's goals are in alignment, as in multiteam systems (MTSs), it becomes nearly impossible to achieve the MTS's goal without delay or scheduling conflicts. Although operating with an internal team leader works fine for the complicated domain, a shared leadership model is more effective when operating in complex domains.

---

3   Multiteam systems (MTS) are covered in more detail in Part IV, Team Science, Chapter 12, Multiteam Systems.

TEAM LEADERSHIP VARIATIONS

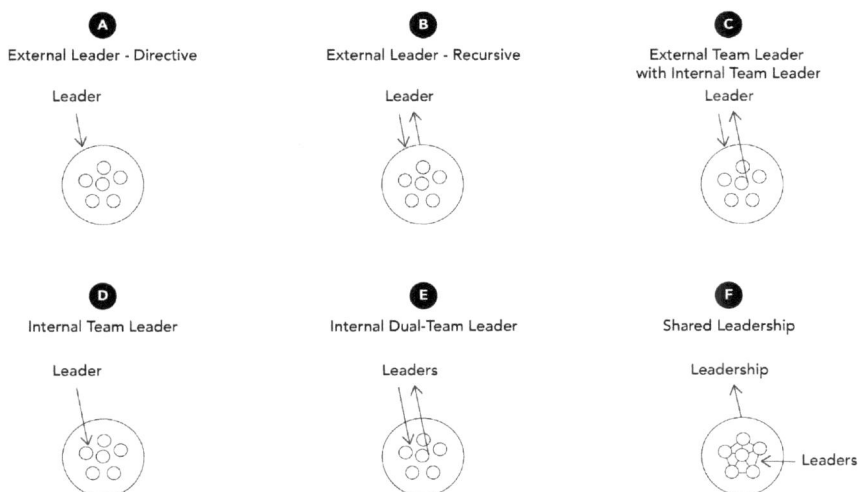

FIGURE 8.2.   Team Leadership Variation Models

Shared leadership, Figure 8.2(f) occurs when all team members participate in the leadership role of the team. Each member utilizes their own strengths, knowledge, and skills for the betterment of the team. This type of model also includes team members taking a step back when other team members have more experience for the task at hand.

This alternating leadership structure results in a model similar to the Team Emergence Leadership Development and Evaluation (TELDE) model (Turner and Baker, 2017), which was described in Chapter 7, and similar to Pearce and Conger's depiction (2003) of shared leadership. This shared leadership model also provides a safe environment that produces positive outcomes, not only for each team member but also for the team as a collective (Wang et al., 2017). Additionally, shared leadership is beneficial in that all team members have an eye, not only on the team's goals but also on the overarching organizational goal. This is critical to success in MTSs in that teams need to be cognizant of their own team goals and the MTS's goals at the same time.

The following section discusses shared leadership in more detail as this type of team leadership is desirable for complex environments and is the team leadership model applied in The Flow System (TFS).

Shared Leadership

Shared leadership can be defined as follows: "Shared leadership reflects a situation where multiple team members engage in leadership and is characterized by collaborative decision-making and shared responsibility for outcomes" (Hoch, 2013: 161). Shared leadership has been shown to be more effective with team processes than traditional and vertical leadership theories (e.g., team effectiveness, team performance, and potency). Conceptually, this makes sense because each team member is more involved with the decision-making processes of the team, team members are more involved, and hence, they become more engaged. This level of engagement also produces an environment in which team members understand the reasoning for the team's course of action and its implementation toward resolving the problem or task at hand (Bergman et al., 2012). Another seminal definition of shared leadership is as follows:

> A dynamic, interactive influence process among individuals in groups for which the objective is to lead one another to the achievement of group or organizational goals or both. This influence process often involves peer, or lateral, influence and at other times involves upward or downward hierarchical influence. (Pearce and Conger, 2003: 1)

Shared leadership has been identified as the optimal model of leadership when the knowledge characteristics of interdependence, creativity, and complexity (Pearce, 2004) are encountered. It is a dynamic process that requires team members to interact while roles and responsibilities shift (self-organizing) in reaction to external perturbations[4] (adaptive). Shared leadership exhibits the characteristics of a complex adaptive system (CAS) as presented in Part II, Complexity Thinking, Chapter 6, Systems Versus Complexity Thinking. To operate, manage, or lead in a complex adaptive system, one needs a leadership process that is dynamic, resilient, and adaptive. Shared leadership meets these requirements and is one of the main reasons that we focus on this type of leadership at the team and group levels of analysis.

Advantages of Shared Leadership

Shared leadership is a model that works well in complex environments: "As the complexity of knowledge work increases, the need for shared leadership also increases: The more complex the task, the lower the likelihood that any one individual can be an expert on all task components" (Pearce, 2004: 49). Research supports shared leadership as a predictor of team outcomes for project-based teams and decision-making teams (Bergman et al., 2012). Shared leadership holds the potential to provide organizations with a competitive advantage because it offers a process for functioning and sharing of information in complex environments (Carson et al., 2007). Shared leadership, as an organizational competitive advantage, affords an organization the ability to address complex

---

4   Perturbations refer to disturbances to current system states. These disturbances can come from internal or external forces.

challenges by being adaptive, by learning, and by utilizing sensemaking tools (Lindsay et al., 2011). Research supports the idea that shared leadership, in which team members experience different leadership styles and multiple leadership experiences, can result in more productive team processes (e.g., lower destructive conflict, team consensus) and more positive emergent states (e.g., trust, cohesion) than teams that did not practice shared leadership (Bergman et al., 2012). Shared leadership aids in providing a competitive advantage in that it results in maximizing a team's resources, coordination, and efficiency (Carson et al., 2007).

Degree of Sharedness

Shared leadership involves both higher-level leaders (executives) and a team's unofficial leader to be fully engaged with the team's progress. This style of leadership is self-organizing because no one individual is directing the team's activities. Rather, the team's membership decides on their course of action and direction. The level of influence by any one team member varies depending on the problem, context of the problem, and the team members' knowledge and experience related to the task at hand. The advantage that shared leadership has compared with other leadership theories is that "the responsibilities of leading are shared among the team members" (Wang et al., 2017: 157). Leadership is a shared construct and varies from one subtask to the next; hence, shared leadership varies in its *degree of sharedness*: "All leadership is shared leadership; it is simply a matter of degree-sometimes it is shared completely while at other times it is not shared at all" (Pearce et al., 2014: 276).

Distributed Coordinated Versus Distributed Fragmented Models

This degree of sharedness is a result of two primary classifications of leadership in shared leadership models:

- The distributed-coordinated model.
- The distributed-fragmented model. (McIntyre and Foti, 2013)

The distributed-coordinated model involves team members who view leadership as being a shared responsibility in which the leadership role is constantly being negotiated (McIntyre and Foti, 2013) among team members based on the task at hand, the team members' knowledge, experience, and external forces experienced by the team.

The distributed-fragmented model involves some team members who identify one member as their informal leader while other team members identify a different member as their informal leader. This fragmented leadership model results in a team with subdivided leadership roles in which the assigned informal leaders may, or may not, communicate with one another. The distributed-fragmented model is often less successful than the distributed-coordinated model of shared leadership. Assignment of the team's leader, during each successive stage, is critical and must be completely agreed on by all team members. For dealing with complexity, the distributed-coordinated model is considered the most effective model because of its flexibility.

Shared Leadership

Three essential characteristics for effective shared leadership to occur have been identified as follows

- low power distance,
- high psychological safety, and
- a strong learning orientation (Lindsay et al., 2011: 534).

Cox et al. also expressed these requirements in their description of the conditions necessary for shared leadership to emerge over time:

> First, team members must understand that constructive lateral influence is a standing performance expectation. Second, members must accept responsibility for providing and responding appropriately to constructive leadership from their peers. Third, the team members must develop skills as effective leadership and followers. Shared leadership, then, is fully expressed only when team members are prepared to function as savvy agents and targets of lateral influence. (Cox et al., 2003: 53)

Low power distance refers to the peer-to-peer influences that take place in shared leadership models. Although there are still some higher hierarchical levels above the team, shared leadership reduces the number of hierarchical power levels and places most of the authority to the team level. This results in the requirement of the lateral influences, peer-to-peer, to remain constructive at all time. Any power that a team member may have external of the team is essentially irrelevant, all team members have equal power inside the team. Remaining constructive within a team setting requires an environment that is psychologically safe for all team members to voice their opinions and concerns, without being ridiculed or reprimanded for speaking out or challenging the status quo. Psychological safety is described as follows:

> A belief that neither the formal nor informal consequences of interpersonal risks, like asking for help or admitting a failure, will be punitive. In psychologically safe environments, people believe that if they make a mistake or ask for help, others will not react badly. Instead, candor is both allowed and expected. Psychological safety exists when people feel their workplace is an environment where they can speak up, offer ideas, and ask questions without fear of being punished or embarrassed. (Edmondson, 2019: 15; see also Edmondson, 1999)

The third requirement relates to having a strong learning environment (Lindsay et al., 2011), or developing team members skills and abilities (Cox et al., 2003). Addressing complex problems deals with unknown-unknowns, which requires teams to learn as collectives. Teams practice complexity thinking and utilize sensemaking tools to help them discover the unknowns, requiring them to learn as a collective rather than individually. This shared learning, in which team members learn individually as well as from other team members through discussion and interactions, results in a shared cognition that emerges in complex adaptive systems. Complex adaptive systems (e.g., teams operating in complexity) are required to learn to be adaptive and to be able to emerge into a new entity that is capable of addressing complexity. Shared leadership is not only a process in which the sharing of leadership roles and responsibilities occurs, but also in which team members learn and mentor one another through the process.

A Few Models of Shared Leadership

New Product Development Teams

In their model of shared leadership for new product development teams, Cox et al. (2003) highlighted the following relevant leadership types that team members should be cognizant of or trained in: transactional and transformational leadership, directive leadership, and empowering leadership. Transformational leadership is composed of four core components:

- idealized influence,
- inspirational motivation,
- intellectual stimulation, and
- individualized consideration (Bass, 1985; see also Thomas, 2017).

Transformational leadership "adopts a more symbolic emphasis on commitment, emotional engagement, or fulfillment of higher-order needs such as meaningful professional impact or desires to engage in breakthrough achievements" (Cox et al., 2003: 56). Transactional and transformational leadership and characteristic definitions are outlined in Table 8.1. In contrast, transactional leadership refers to leading by using rewards and punishment techniques, such as material rewards, affective praise, and peer-pressure (Cox et al., 2003). These leadership styles have been included in a comprehensive leadership theory called the full-range leadership theory (Bass, 1985), which includes transactional, transformation, and laissez-faire leadership. Laissez-faire leadership involves that act of doing nothing, referring to the point at which a leader sometimes needs to simply stand by and do nothing. The idea is that leadership will vary depending on the contextual setting: at times, transactional leadership is required, whereas other times, transformational leadership will be required. In relation to shared leadership, transactional leadership has been shown to be more effective in stable

TABLE 8.1.   Transactional and Transformational Leadership and Characteristics

Theory/Characteristic	Definition
Transactional Leadership	Provides valued rewards and praise, compensation, or other desirable outcomes that are contingent on responsible behavior (Bass, 1985; Pearce et al., 2014).
Transformation Leadership	Foster inspiration and commitment to an overarching responsible vision or mission (Pearce et al., 2014; Bass, 1985).
Idealized influence	Exhibiting confidence and charisma that arouse strong emotions and loyalty from followers (Thomas, 2017: 386).
Inspirational motivation	Articulating organizational goals, communicating high expectations, and convincing followers of the importance of these goals (Thomas, 2017: 386–387).
Intellectual stimulation	Encouraging innovative ways of thinking and doing things and breaking away from existing routines and norms (Thomas, 2017: 387).
Individualized consideration	Attending to the individual needs of followers, acting as their coach, and listening to their concerns (Thomas, 2017: 387).

environments in which innovation is only incremental. In contrast, transformational leadership helps to improve innovation and effectiveness (Cox et al., 2003).

Directive leadership is closely connected with providing task-specific guidance and recommendations (Cox et al., 2003), instructions that guide efforts toward responsible goals (Pearce et al., 2014), which aligns nicely with team-based systems. Directive leadership has been shown to provide structure for complex tasks and to improve team performance (Cox et al., 2003).

Empowering leadership focuses on how people engage in responsible influence and self-influence (Pearce et al., 2014) and identifies with such concepts as self-empowerment and self-leadership (Cox et al., 2003). Empowering leadership promotes vertical and horizontal influence, as required in team settings, by fostering "peer encouragement and support of self-goal setting, self-evaluation, self-reward, and self-development" (Cox et al., 2003: 57) and places the decision-making capabilities in the hands of the team members. Table 8.2 defines and describes the behaviors necessary for directive leadership and empowering leadership.

In this shared leadership model,[5] leadership among the team members is practiced using one of these three leadership styles: transactional and transformational leadership, directive leadership, and empowering leadership. Oversight of the team

---

5   This example of shared leadership in the contextual setting of new product development teams has its theoretical foundation in the full range theory of leadership (transformational, transactional, laissez-faire), directive leadership, and empowering leadership. We present these as leadership theories, leadership models, or leadership styles. There are distinctive differences among these theories, models, and styles, but for descriptive purposes, we use these terms interchangeably.

TABLE 8.2.   Directive and Empowering Leadership

Theory	Definition/Description
Directive Leadership	Leadership primarily relies on position power. Involves planning and organizing roles and responsibilities for subordinates. Behaviors include (a) issuing instructions and commands, and (b) assigning goals. (Pearce and Sims, 2002: 173–174)
Empowering Leadership	Emphasizes the development of follower self-management or self-leadership skills. Leading others to lead themselves. Behaviors include (a) encouraging independent action, (b) encouraging opportunity thinking, (c) encouraging teamwork, (d) encouraging self-development, (e) using participative goal setting, and (f) encouraging self-reward. (Pearce and Sims, 2002: 175)

comes from vertical leadership (outside of the team) in which an external leader (project manager for new product development teams) provides guidance, support, and resources for the team. This vertical leadership functions four essential roles: team formation, boundary management, leadership support, and shared leadership maintenance (Cox et al., 2003). Vertical leadership begins with the original makeup of the team and continues adding, or removing, team members as the project evolves on an as-needed basis. Boundary management is critical in that it buffers the team from external forces, allowing the team to concentrate on its taskwork. This role also provides resources for the team and makes any necessary external connections when needed. Leadership support involves maintaining a positive shared leadership climate at all times. This function also involves shared leadership management in which training and encouragement is provided.

This model of shared leadership provides specific team characteristics; proximity, team size, ability (or knowledge, skills, and abilities [KSAs]), team diversity, and team maturity (Cox et al., 2003). Moderating variables also are provided, which include interdependence and complexity along with outcome variables of team responses and team effectiveness (Cox et al., 2003). At this point, we will not define each of these team characteristics, moderating variables, or outcome variables. What is important, however, is that they all relate to team-related functions. This point is stressed in military leadership development efforts: "The resulting level of teamwork that is developed is instrumental in shaping overall team learning capacity, which, in turn, is positively related to team leadership capacity (primarily in the form of shared leadership)" (Lindsay et al., 2011: 533). This view highlights the importance of utilizing the knowledge from the field of team science to provide fully functional and effective teams. The leadership practiced within each team, among the team members, and the leadership that facilitates and guides the team (boundary management) is important. It is equally important, however, to ensure that each team is functioning as a team. Part IV, Team Science, Chapter 10, Teams, Teamwork, and Taskwork, provides more detail on the topic.

Military Models and Team Coaching

Shared leadership has been viewed as a team construct consisting of the dimensions of shared purpose, social support, and voice (Lindsay et al., 2011). Shared purpose involves the team's understanding and willingness to achieve the stated goals, social support relates to emotional and psychological support (psychological safety), and voice relates to assuring everyone that they will be heard (psychological safety) (Lindsay et al., 2011). Rather than utilizing vertical leadership, as in the case with the Cox et al. (2003) shared leadership model, team coaching is recognized.

Team coaching provides the level of external leadership required for shared leadership with a focus on supportive and reinforcing support for the team members (Lindsay et al., 2011). For the military, when the environment changes from normal to chaotic, shared leadership was found to be the best leadership model. Leadership, as a team construct, holds things together in times of emergency and chaos (Lindsay et al., 2011). From experience with military missions and emergency operations, shared leadership has been found to be have the following characteristics:

> (a) embrace continuous learning particularly because dangerous situations demand it, (b) are willing to share risks, (c) share a common lifestyle and emphasize shared values rather than material possessions, (d) readily learn competencies that allow them to make rapid and effective decisions, (e) deliberately create real leadership feelings of trust and care, and (f) exhibit and deserve team loyalty. (Lindsay et al., 2011: 546; see also Kolditz, 2007)

These characteristics touch on both leadership and teamwork characteristics that are similar to the characteristics of transformational and transactional, directive, and empowering leadership. They have an added emphasis on team characteristics, as presented in the Cox et al. (2003) shared leadership model.

Extreme Teaming and Functional Leadership

Edmondson and Harvey (2018) used functional leadership, rather than vertical leadership, for extreme teams.[6] Functional leadership focuses on the leader–team interactions rather than the leader–follower interactions as in most leadership theories (Edmondson and Harvey, 2017). Table 8.3 defines and describes the behaviors of functional leadership.

Functional leadership for extreme teaming was presented as a two-by-two taxonomy, or matrix, that presents four leadership functions: build an engaging

---

6   Extreme teaming is defined as a cross-boundary collaboration that is required in new product development teams and innovative environments as well as in complex environments (Edmondson and Harvey, 2017).

TABLE 8.3. Functional Leadership

Theory	Definition/Description
Functional Leadership	Focuses on leader–team interactions, not leader–follower interactions. Emphasizes necessary conditions for effective team performance. The four leadership functions are as follows: building an engaging vision, empowering agile execution, cultivating psychological safety, and developing shared mental models. (Edmondson and Harvey, 2017: 26–34, 110)

vision, empower agile execution, cultivate psychological safety, and develop shared mental models (Edmondson and Harvey, 2017). Across the first row of the taxonomy, building an engaging vision is accomplished by leaders focusing on interpersonal and motivational factors, whereas empowering agile execution is accomplished through a leader's focus on technical (KSAs) and motivational factors (Edmondson and Harvey, 2017). Across the second row of the taxonomy, cultivating psychological safety is achieved by focusing on interpersonal factors and facilitation or removing barriers, whereas developing shared mental models is attained through concentration on technical and facilitating factors (Edmondson and Harvey, 2017). In sum, this taxonomy is described in the following description:

> The four leadership functions fall into two sets of categories. First, they address either an interpersonal or a technical challenge, and second, they serve primarily to motivate effort or to facilitate progress. Each category finds support in-and adds to-prior research on team effectiveness and on knowledge in organizations. (Edmondson and Harvey, 2017: 131)

Other functional leadership models view leadership outcomes as followers' states, team states, emergence, and effectiveness through the antecedents (inputs) of a leader's demands and affordances (Zaccaro et al., 2018). Leadership demands relate to a leader's perception of the contextual or situational factors. A leader's perception of the situation affects how that leader reacts and will influence the strategies that a leader selects. Leadership affordances relate to a leader recognizing opportunities based on the environmental conditions, that is, the contextual factors (Zaccaro et al., 2018).

These contextual or situational factors are also influenced through the level of task complexity and social complexity present at any given time. Task complexity can be broken down into three categories: "component complexity (number of acts and information cues involved), coordinative complexity (type and number of relationships among acts and cues), and dynamic complexity (changes in acts and cues and the relationship among them)" (Wood et al., 1987: 418). Likewise, as Zaccaro et al. (2018) explain, social complexity relates to the frequency of the different social situations, the number and diversity of units and stakeholders, and the different roles required of the leader for each setting.

Functional leadership leads to three key considerations when viewing leadership at the individual and team levels:

- leaders choose, shape, or react to situations' perceived performance requirements and leadership requirements;
- situations cue particular individual differences for the expression of functional leadership behaviors; and
- situations afford certain leadership choices and actions that are, in turn, derived from individual differences. (Zaccaro et al., 2018: 34)

Leadership is contextual, is bounded upon performance expectations, exhibits a power-distance relationship, and result in KSAs and experiences. Functional leadership, as an external team leadership model, needs to focus on the contextual team setting, to come to a shared understanding of performance expectations for the leader and for the team, to share responsibilities to reduce any power-distance conflicts, and to share roles and responsibilities to benefit from each members' KSAs and experiences.

Conclusion

The main takeaway from this chapter is that "shared leadership can be developed, and it is generally a good thing in terms of enhancing team effectiveness" (Lindsay et al., 2011: 535). For organizations planning to transition to team-based structures, or for organizations that already have transitioned with little to no success, structuring your teams based on one of the shared leadership models could be beneficial. The different views and models of shared leadership, along with the various modes of providing external leadership for shared leadership (vertical leadership, boundary management, team coaching, functional leadership) provide organizations with options, not a prescription. This is not a one-size-fits-all solution as each organization differs in their membership, expertise, resources, and level of complexity in which they are operating.

Google provides an example of lessons learned from an organization that has been operating using team-based structures for some time. In identifying which behaviors a manager should encompass, Google set out to collect data and identify the key characteristics that made managers great. Their results identified the following 10 behaviors. A good manager,

1. Is a good coach.
2. Empowers [the] team and does not micromanage.
3. Creates an inclusive team environment, showing concern for success and well-being.
4. Is productive and results-oriented.
5. Is a good communicator—listens and shares information.
6. Supports career development and discusses performance.

7. Has a clear vision/strategy for the team.
8. Has key technical skills to help advise the team.
9. Collaborates across Google.
10. Is a strong decision maker. (re:Work, n.d.)

In viewing these management behaviors, one would find similarities to the previous leadership options that we proposed in this chapter (vertical leadership, boundary management, team coaching, functional leadership). These characteristics are more similar to external team leadership characteristics than they are to managerial characteristics. Notably, this list of managerial behaviors has a greater focus on teams and teamwork than traditional managerial functions (compare with list provided in Chapter 7, A Word About Leadership, in the section "Leadership Versus Manager"). This is one example that shows how the leadership functions must be altered to being team focused when changing to team-based structures.

Organizations must realize that they should be building the organizational structure that they need for the future rather than adding team-based structures to an existing organizational structure. This point is highlighted by McCord in her discussion about the transformation experienced at Netflix: "You've got to hire *now* the team you wish to have in the future" (McCord, 2017: 72). This point can be extended beyond just hiring a team for the future to identifying what is required to build the organizational and leadership structure that will be needed for the future. These different models of shared leadership and external team leadership can guide organizations as they move forward.

Leadership beyond an organization's team-structure requires a different type of leadership. The next chapter presents some of the leadership models that have been shown to work well in planning for a future of disruption and complexity. These models also work well with team-based structures.

References

Bass BM. (1985) *Leadership and performance beyond expectations.* New York, NY: The Free Press.

Bergman JZ, Rentsch JR, Small EE, et al. (2012) The shared leadership process in decision-making teams. *Journal of Social Psychology* 152: 17–42.

Carson JB, Tesluk PE and Marrone JA. (2007) Shared leadership in teams: An investigation of antecedent conditions and performance. *Academy of Management Journal* 50: 1217–1234.

Cox JF, Pearce CL and Perry ML. (2003) Toward a model of shared leadership and distributed influence in the innovation process: How shared leadership can enhance new product development team dynamics and effectiveness. In: Pearce CL and Conger JA (eds) *Shared leadership: Reframing the hows and whys of leadership.* Thousand Oaks, CA: Sage, 48–76.

Edmondson A. (1999) Psychological safety and learning behavior in work teams. *Administrative Science Quarterly* 44: 350–383.

Edmondson AC. (2019) *The fearless organization: Creating psychological safety in the workplace for learning, innovation, and growth.* Hoboken, NJ: Wiley.

Edmondson AC and Harvey J-F. (2017) *Extreme teaming: Lessons in complex, cross-sector leadership.* Bingley, UK: Emerald Publishing.

Folkestad J and Gonzalez R. (2010) Teamwork for innovation: A content analysis of the highly read and highly cited literature on innovation. *Advances in Developing Human Resources* 12: 115–136.

Hoch JE. (2013) Shared Leadership and Innovation: The Role of Vertical Leadership and Employee Integrity. *Journal of Business and Psychology* 28: 159–174.

Kolditz TA. (2007) *In extremis leadership: Leading as if your life depended on it.* San Francisco, CA: Jossey-Bass.

Lindsay DR, Day DV and Halpin SM. (2011) Shared leadership in the military: Reality, possibility, or pipedream? *Military Psychology* 23: 528–549.

Manderscheid SV. (2008) New leader assimilation: An intervention for leaders in transition. *Advances in Developing Human Resources* 10: 686–702.

McCord P. (2017) *Powerful: Building a culture of freedom and responsibility.* San Francisco, CA: Silicon Guild.

McIntyre HH and Foti RJ. (2013) The impact of shared leadership on teamwork mental models and performance in self-directed teams. *Group Processes and Intergroup Relations* 16: 46–57.

Pearce CL. (2004) The future of leadership: Combining vertical and shared leadership to transform knowledge work. *Academy of Management Executive* 18: 47–57.

Pearce CL and Conger JA. (2003) All those years ago: The historical underpinnings of shared leadership. In: Pearce CL and Conger JA (eds) *Shared leadership: Reframing the hows and whys of leadership.* Thousand Oaks, CA: Sage, 1–18.

Pearce CL and Sims HP. (2002) Vertical versus shared leadership as predictors of the effectiveness of change management teams: An examination of aversive, directive, transactional, transformational, and empowering leader behaviors. *Group Dynamics: Theory Research and Practice* 6: 172–197.

Pearce CL, Wassenaar CL and Manz CC. (2014) Is shared leadership the key to responsible leadership? *Academy of Management Perspectives* 28: 275–288.

re:Work. (n.d.) *Learn about Google's manager research.* Available at: https://rework.withgoogle.com/guides/managers-identify-what-makes-a-great-manager/steps/learn-about-googles-manager-research/.

Thomas NWH. (2017) Transformational leadership and performance outcomes: Analysis of multiple mediation pathways. *The Leadership Quarterly* 28: 385–417.

Turner JR and Baker R. (2017) Team emergence leadership development and evaluation: A theoretical model using complexity theory. *Journal of Information and Knowledge Management* 16: 17.

Uhl-Bien M and Marion R. (2009) Complexity leadership in bureaucratic forms of organizing: A meso model. *The Leadership Quarterly* 20: 631–650.

Uhl-Bien M, Marion R, and McKelvey B. (2007) Complexity leadership theory: Shifting leadership from the industrial age to the knowledge era. *The Leadership Quarterly* 18: 298–318.

van Knippenberg D. (2017) Team leadership. In: Salas E, Rico R, and Passmore J (eds) *The Wiley Blackwell handbook of the psychology of team working and collaborative processes.* Malden, MA: Wiley Blackwell, 345–368.

Wang L, Wan J, Liu Z, et al. (2017) Shared leadership and team effectiveness: The examination of LMX differentiation and servant leadership on the emergence and consequences of shared leadership. *Human Performance* 30: 155–168.

Wood RE, Mento AJ, and Locke EA. (1987) Task complexity as a moderator of goal effects: A meta-analysis. *Journal of Applied Psychology* 72: 416–425.

Zaccaro SJ, Green JP, Dubrow S, et al. (2018) Leader individual differences, situational parameters, and leadership outcomes: A comprehensive review and integration. *The Leadership Quarterly* 29: 2–43.

CHAPTER 9

Strategic, Instrumental, and Global Leadership

To lead and manage team-based structures and newly integrated leadership styles presented in the previous chapters, new or modified leadership styles are necessary at the executive levels as well. The amount of change required in existing leadership practices at the executive levels will vary from one organization to the next. Regardless of the gap between what is needed and what currently exists, leadership at the executive levels needs to be addressed so that leaders are better able to not only manage team-based structures and leadership styles but also to lead in disruptive economies and in complex environments.

Leadership at the executive level must be changed to meet the needs of the future, not the present. Leadership becomes a holistic organizational entity that flows from the executive level to the external leadership ranks for the shared leadership model (vertical leadership, boundary management, team coaching, functional leadership) to the team level as well as flowing in the opposite direction equally. This holistic organizational model is presented in The Flow System (TFS) as distributed leadership and is depicted in Figure 9.1. In this holistic model, leadership becomes "a growing, learning, adapting living organism that is in constant flux to identify and develop new opportunities that add value for customers" (Denning, 2018: 20). With the understanding that support and change begin with leadership at the executive levels, the transformation required to design, structure, and support the new team-based and leadership structures comes from the C-suite or executive group. This change and transformation, however, is required not only external to the executive suite but internal as well. In most cases, current leadership practices will not provide enough support and autonomy for team-based structures and leadership to be successful; therefore, executive levels also are required to make some changes. This chapter provides

New Leadership Structure
(Required for Team-Based Systems)

Integrated leadership Theory:
Strategic Leadership, Instrumental Leadership,
Global Leadership

Executive Level

Blended — — — ↓ — — ↓ — — ↓ — — — Structure

Vertical Leadership, Boundary Management,
Team Coaching, Functional Leadership

Team-Based System

Shared Leadership

DISTRIBUTED LEADERSHIP

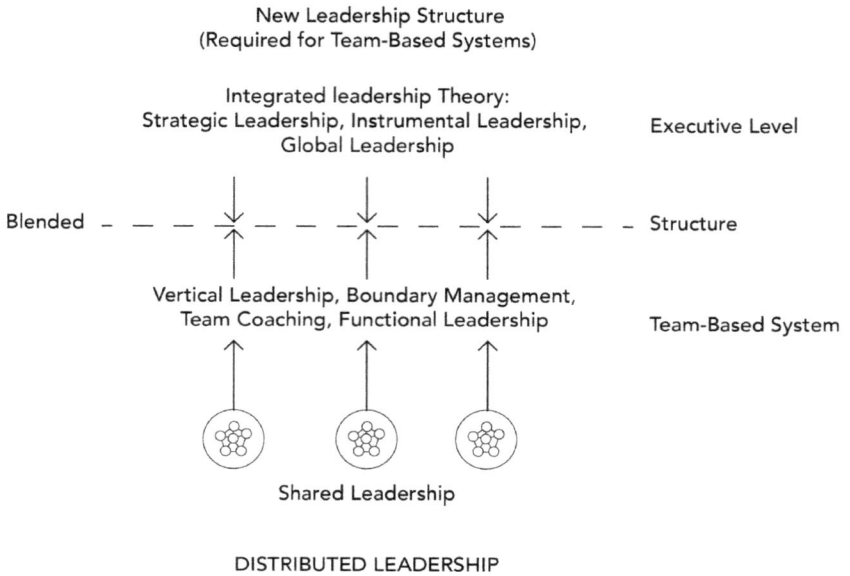

FIGURE 9.1.   Distributed Leadership

examples of a few leadership theories that could meet this need for change at the executive levels, including strategic, instrumental, and global leadership theories as potential models for top-level leaders.

These leadership theories (i.e., strategic, instrumental, and global) are current leadership theories that have been shown to work well with team-based structures and that also are geared toward complexity, creativity, and innovation as opposed to just managing the status quo. Current trends and calls in the discipline of leadership are calling for new leadership theories that integrate existing theories with newer theories to meet the needs of the global economy and the new customer better. These calls are for integrated theories that combine what already is known to meet the demands required of new contextual situations with the development of additional theories that cut across the lower levels (micro) to the higher levels (macro) of leadership within an organization (Crossan et al., 2008; Waldman et al., 2004; Waldman and Yammarino, 1999). The leadership theories present at the macro levels of an organization must work well with the leadership theories at the micro levels, thus providing a seamless flow of leadership across the whole organization. The strategic, instrumental, and global leadership models discussed in the current chapter have been shown to work well with other team-based leadership theories as presented in the previous chapter, Chapter 8, Team and Distributed Leadership. The integration of these macro- and micro-level leadership theories results in our description of distributed leadership for TFS.

Strategic Leadership

Strategic leadership has its roots in the upper-echelon theory (Hambrick and Mason, 1984). The upper-echelon theory views an organization's success as a factor of leaders at the top of that organization. This theory views the "upper echelon characteristics as determinants of strategic voices and, through these choices, of organizational performance" (Hambrick and Mason, 1984: 197). Examining individuals who determine an organization's strategic decisions, who develop its purpose and direction, and who become the face of the organization (Hernandez et al., 2011) could be perceived as being shortsighted.

Because leaders at the executive levels of any organization do not act in isolation, research has expanded to look at top management teams as the primary focus as opposed to looking at just one individual executive. This focus on the top management teams provides "stronger explanations for organizational outcomes than a focus on one or a few individuals only" (Hernandez et al., 2011: 1179). Because of this executive team perspective, strategic leadership has been classified as an executive leadership theory (Dionne et al., 2014). This new collective view of leadership had resulted in strategic leadership research viewing behaviors such as the following: making strategic decisions, communicating a vision, developing tomorrow's leaders, and developing organizational structures, to name but a few (Boal and Hooijberg, 2000).

As organizations face increasing levels of complexity, leadership at the top also deal with more complexity than in previous generations. Today, this leadership at the top is composed mostly of executive teams and boards. It makes sense that the leadership model at the top executive levels should be capable of dealing with, and operating in, complex environments. This is especially true when the lower levels of the organization also are dealing with these same complex problems.

Strategic leadership is one theory that views executive leadership within the context of "ambiguity, complexity, and informational overload" (Boal and Hooijberg, 2000: 516). More recently, strategic leadership has viewed leadership as a complex adaptive system, providing new perspectives in leadership that include "issues of shared, distributed, collective, relational, dynamic, emergent and adaptive leadership processes" (Uhl-Bien and Marion, 2009: 631). Leadership in complex adaptive systems are more focused on creating a structure that provides the context, promotes the coordination, guides the interactions required among agents, and channels knowledge and vision to achieve an organization's goals (Boal and Schultz, 2007). Leaders, in complex adaptive systems, and through strategic leadership, provide *leadership of organizations* rather than providing *leadership in organizations* as newer and traditional leadership theories provide (Boal and Schultz, 2007; Dubin, 1979). Here, strategic leadership "focuses on the creation of meaning and purpose for the organization" (Boal and Schultz, 2007: 412). The following description of strategic leadership is provided by Boal and Schultz (2004), this description further illustrates strategic leaderships' role in complex adaptive systems and its relevance and place in TFS as a model of leadership for the executive levels:

> Strategic leadership is a series of decisions and activities, both process-oriented and substantive in nature, through which, over time, the past, the present, and the future of the organization coalesce. Strategic leadership forges a bridge between the past, the present and the future, by reaffirming core values and identity to ensure continuity and integrity as the organization struggles with known and unknown realities and possibilities. Strategic leadership develops, focuses, and enables an organization's structural, human, and social capital and capabilities to meet real-time opportunities and threats. Finally, strategic leadership makes sense of and gives meaning to environmental turbulence and ambiguity, and provides a vision and road map that allows an organization to evolve and innovate. (as cited in Boal and Schultz, 2007: 412)

Components of Strategic Leadership

Earlier conceptualizations of strategic leadership consisted of the components of *absorptive capacity, adaptive capacity,* and *managerial wisdom* (Boal and Hooijberg, 2000). These components of strategic leadership are listed in Table 9.1. Absorptive capacity deals with one's ability to learn. At the collective level, that of the executive team level, learning is achieved through mutual cooperation, trust, support, active listening, and information exchange (Boal and Hooijberg, 2000). Adaptive capacity relates to being adaptive to change, to address disequilibrium and hypercompetitive markets (Boal and Hooijberg,

TABLE 9.1.   Strategic Leadership Components and Behaviors

Component of Strategic Leadership	Description/Behaviors
Absorptive Capacity	The ability to learn, to recognize new information, assimilate it, and apply it toward new ends. Creates an organizational context in which learning takes place. Encourages plausible judgement, active learning, periodic information exchange, and working consensus.
Adaptive Capacity	The ability to change. Provides strategic flexibility allowing an organization to protect or respond quickly to changing competitive conditions. Leaders have cognitive and behavior complexity and flexibility coupled with an openness to and acceptance of change.
Managerial Wisdom	The ability to perceive variation in the environment and an understanding of the social actors and their relationships. The capacity to take the right action at a critical moment. Behaviors include social intelligence, interpersonal intelligence, social awareness, and social skills.

Source: Boal and Hooijberg (2000: 517–518).

2000). Adaptive capacity is similar to the characteristics of complex adaptive systems—that is, adaptable while operating between order and chaos. Last, managerial wisdom relates to one's ability to view the environment and to understand the variations in social complexities (Boal and Hooijberg, 2000). Managerial wisdom is similar to the practice of sensemaking that was discussed in Part II, Complexity Thinking.

In providing an integrated theory of strategic leadership, Boal and Hooijberg (2000) combined strategic leadership with charismatic, transformational, and visionary leadership. This integration included optimizing current knowledge from existing leadership theories while also including more context related to research on teams (Boal and Hooijberg, 2000). A leader's charisma provides vision and also instills crisis management when needed (Boal and Hooijberg, 2000); transformational leadership provides intellectual stimulation, individual consideration, and inspiration (Bass, 1985); and visionary leadership provides reason for change that combines the past, present, and future of the organization (Gioia and Thomas, 1996). In viewing strategic leadership, it is essential to ensure that the executive team's strategy aligns with their sensemaking activities; that their executive team's identity perception aligns with the organizations; that their strategy is aligned with, and perceived by, other organizational agents and relevant to the complex environment these agents are operating in (Gioia and Thomas, 1996); and that the strategy is aligned with customer needs. These alignment issues addressed by strategic leadership, ultimately, when in order, result in flow by maximizing value to the customer.

One essential tool for strategic leadership is having storytelling ability. Storytelling is a process of sensemaking and knowledge sharing that involves dialogue and narratives. As storytelling combs through narratives from multiple agents, patterns are uncovered to show new knowledge behind previous unknowns, and this new knowledge begins to uncover a story, beginning the transition from the unknown to the known. Storytelling also allows individuals, under the guide of the strategic leader, to "share their explicit knowledge and their implicit understandings" (Boal and Schultz, 2007: 427). Because leadership is a collective process, overseeing team-based structures, storytelling helps to extend a strategic leaders' knowledge and vision to teams by producing a shared schemas or shared understanding among team members. These shared schemas evolve over time, arise from the interactions among team members, and are inherent in the structures provided by strategic leadership by the team-based structures and the teamwork and psychological safety afforded by such structures. Storytelling as a leaders' sensemaking tool consists of a reciprocal relationship between telling and listening (Nossel, 2018). Through storytelling, strategic leaders are better able to "articulate their visions by telling stories and promoting dialogue in which an organization's past, present, and future coalesce: stories and dialogue about our history; stories and dialogue about who we are; stories and dialogue about who we can become" (Boal and Schultz, 2007: 426).

Integrated Strategic Leadership Theories

In viewing strategic leadership in today's dynamic environment, Crossan et al. (2008) integrated leadership theories through transcendent leadership. Transcendent leadership describes one who "leads within and amongst the levels of self, others, and organization" (Crossan et al., 2008: 570). A figure presented by Crossan et al. included a Venn diagram in which three circles were drawn, one each for leadership and self, leadership of others, and leadership of organization. At the intersecting area for these three circles, where all three overlap in the center, is transcendent leadership. Transcendent leadership fills in the gap between current leadership practices of self, others, and the organization. Transcendent leadership, in this integrated model of strategic leadership, is intended to produce leaders who can "transcend the levels, as it captures the quality of going above and beyond, within and between levels" (Crossan et al., 2008: 576). This type of strategic leadership would work well in coordinating activities from the team-based leadership models of vertical leadership, boundary management, team coaching, and functional leadership (Figure 9.1).

In bringing the CEO's characteristics to the forefront, Waldman et al. (2004) highlighted a model of charismatic leadership at the strategic level. This model presents a CEOs charisma in conjunction with the strategic change and any perceived environmental uncertainty as factors that are associated with an organizations' performance. Charisma is described as a relationship that a leader has with one or more followers that are perceived as being favorable by the follower or followers.

Charismatic behaviors may include "providing a sense of mission, articulating a future-oriented, inspirational vision based on powerful imagery, values, and beliefs. . . . Showing determination when accomplishing goals and communicating high performance expectations" (Waldman et al., 2004: 358).

Charismatic leadership has been shown, in recent research, to be linked positively with strategic leadership. For example, charismatic leadership stimulated cohesion among group members that identify with the charismatic leader's behaviors, vision, sagas, and storytelling (Waldman and Yammarino, 1999). Charismatic leadership affects both direct and indirect followers while manifesting differently at various organizational levels (Waldman and Yammarino, 1999) and while holding promise for influencing the shared leadership ranks present below the blended structure line depicted in Figure 9.1.

Instrumental Leadership

Instrumental leadership is an example of an integrated leadership theory that utilizes functional leadership along with the full range leadership model presented by Bass (1985), which incorporates transactional, transformational, and laissez-faire leadership theories. Instrumental leadership extends the functional and full range leadership models to include both strategic and pragmatic leadership. Instrumental leadership fills in the gaps found in the full-range leadership model. Beyond leading through influence, reward, and punishment techniques, as with transactional and transformational leadership, researchers felt that leadership also must provide direction for organizations to adapt to their external environments and provide solutions to complex problems (Antonakis and House, 2014). Instrumental leadership adds to the full-range leadership model by also including leaders who are capable of "identify[ing] strategic and tactical goals while monitoring team outcomes and the environment" (Antonakis and House, 2014: 747).

The full-range leadership model does not address the growing need for shared leadership and has been identified as a model that operates outside of team activities (Antonakis and House, 2014; Morgeson et al., 2010). Some of the shortcomings of the full-range model include the following:

> (a) strategic structuring and planning (e.g., identifying strategies and goals), (b) providing direction and resources (e.g., classifying tasks, ensuring the team has sufficient resources), (c) monitoring the external environment (e.g., monitoring changes), and (d) monitoring performance and feedback provision (e.g., monitoring individual performance and providing corrective feedback). (Antonakis and House, 2014: 748)

These deficits are captured by integrating strategic and pragmatic leadership with the full-range leadership model to develop instrumental leadership. Transformational and transactional leadership provides tools for a leader to motivate followers to complete their task while offering opportunities for them to advance themselves in the process (transformational). Strategic leadership provides making strategic decisions, communicating a vision, and developing leaders and organizational structures (see the section "Strategic Leadership"). Pragmatic leadership provides a focus on organizational outcomes and problem solving. Instrumental leadership provides a *fuller range* of behaviors than transformational and transactional leadership by integrating strategic and task-monitoring behaviors (Antonakis and House, 2014). By integrating the full-range leadership model with strategic and pragmatic leadership, instrumental leadership can be defined in the following:

> The application of leader expert knowledge on monitoring of the environment and of performance, and the implementation of strategic and tactical solutions. Strategically, leaders monitor the external environment and identify strategies and goals. From a follower work facilitation point of view, leaders provide direction and resources, monitor performance and provide feedback. (Antonakis and House, 2014: 749; see also Morgeson et al., 2010)

Components of Instrumental Leadership

Research has supported the concept that leaders are capable of displaying multiple forms of leadership, resulting in the mixed leader profile that is presented here along with instrumental leadership theory. The following factors for the conceptual model of instrumental leadership are provided in Table 9.2.

TABLE 9.2.   Instrumental Leadership

Instrumental Leadership Factors	*Description*
Transformational Leadership	Proactive, raise follower awareness for transcendent collective interests, and help followers achieve extraordinary goals.[a]
Idealized Influence Attributes	The socialized charisma of the leader.[a]
Idealized Influence Behaviors	The charismatic actions of the leader that are centered on values, beliefs, and a sense of mission. The way leaders energize their followers.[a]
Inspirational Motivation	The ways leaders energize their followers by viewing the future with optimism, stressing ambitious goals, projecting an idealized vision, and communicating to followers that the vision is achievable.[a]
Intellectual Stimulation	Leader actions that appeal to followers' sense of logic and analysis by challenging followers to think creatively and find solutions to difficult problems.[a]
Individualized Consideration	Leader behavior that contributes to follower satisfaction by advising, supporting, and paying attention to the individual needs of followers, and thus allowing them to develop and self-actualize.[a]
Transactional Leadership	An exchange process based on the fulfillment of contractual obligations.[a]
Contingent Reward	Leader behaviors focused on clarifying role and task requirements and providing followers with material or psychological rewards contingent on the fulfillment of contractual obligations.[a]
Management-by-Exception Active	The active vigilance of a leader whose goal is to ensure that standards are met.[a]
Management-by-Exception Passive	Leaders only intervene after noncompliance has occurred or when mistakes have already happened.[a]
Laissez-faire Leadership	The absence of a transaction of sorts with respect to leadership in which the leader avoids making decisions, abdicates responsibility, and does not use their authority.[a]
Strategic Leadership	Focuses on the creation of meaning and purpose for the organization.[b]
Environmental Monitoring	Leader actions regarding scanning the internal and external organizational environments.[c]
Strategy Formulation and Implementation	Leader actions focused on developing policies, goals, and objectives to support the strategic vision and mission.[c]
Pragmatic Leadership	Utilize a problem-solving approach that aims to intellectually stimulate followers through effective communication steeped in logical appeals.[d]
Path-Goal Facilitation	Leader behaviors targeted toward giving direction, support, and resources, removing obstacles for goal attainment and providing path-goal clarifications.[c]
Outcome Monitoring	Entails leader provision of performance-enhancing feedback useful for goal attainment.[c]

Source: [a]Antonakis et al. (2003: 264–265); [b]Boal and Hooijberg (2000: 516); [c]Antonakis and House (2014: 750); and [d]Lovelace et al. (2019: 98).

Although this integrated instrumental leadership theory may seem complicated at first, it provides a more detailed model for leadership at the executive levels for any organization. Instrumental leadership utilizes knowledge gained from the full-range leadership model while also including leadership characteristics that "better explains how leadership happens not only on an interpersonal and transactional level but also on a strategic and work-facilitation level" (Antonakis and House, 2014: 765). Aside from looking inward at the organization and externally at environmental threats, it is also advised to incorporate some component level of global leadership at the executive levels. The following section presents global leadership.

Global Leadership

Globalization has changed the landscape of leadership significantly (Gehrke and Claes, 2017) and will continue to do so for some time. An example of this is present in the following description of global leaders' task: "Global leaders have to connect people across countries and engage them to global team collaboration in order to facilitate complex processes of knowledge sharing across the globe" (Gehrke and Claes, 2017: 373). This description also touches each of the three components of TFS: complexity thinking, distributed leadership, and team science. Globalization has resulted in managers and leaders working full time with people from multiple countries and cultural backgrounds, sometimes within the same organization, sometimes within the same organization but located in different countries, and other times located in different countries and different organizations. In this way, globalization has changed the leadership landscape to being "less linear and now more non-linear in nature" (Bird and Mendenhall, 2016: 119).

As globalization continues to affect more and more countries, industries, and organizations (large and small), techniques and tools to navigate these global markets are needed even more today. Global markets are, by definition, complex environments that also have been called a VUCA (volatile, uncertain, complex, ambiguous) environment with the theme of managing or operating in complexity (Gehrke and Claes, 2017). Existing theories of leadership fail to fully capture the leadership within the context of global diversity (Steers et al., 2012). Global leadership should advance, as a minimum, our understanding of the following processes: "(a) the meaning of leadership as a cultural construct, and (b) the variations in local expectations regarding leader behavior" (Steers et al., 2012: 481).

Research has shown that leaders act according to the desired leadership of the culture in which they are operating (Dorfman et al., 2012), meaning that some bias normally exists in a leader's behavior and in the decisions that are made. As a leader becomes exposed to more cultures, more influences enter the picture, making it harder for any one leader to accommodate multiple cultures.

Utilizing global leadership to better address differing cultures and issues should be part of any leaders' toolbox at the upper executive levels, for any organization.

One tool for leading in a complex global environment is to develop a global mind-set for leaders. A global mind-set needs to be developed in emerging leaders. A global mind-set has been defined as "tolerance of ambiguity, flexibility and cultural adaptability" (Bird and Mendenhall, 2016: 121). Other extensions of a global mind-set include four elements: flexibility, acceptance (related to "cognitive complexity"), curiosity (related to "cosmopolitanism"), and empathy and emotional connection (Henson, 2016).

From a global perspective, leaders must be constantly aware of their social responsibilities to local economies, cultures, and problems. Global leaders must learn to "adapt their leadership style to fit local circumstances in order to achieve corporate objectives" (Steers et al., 2012: 479). Essential at the macro level, the executive level, leaders must be well versed and trained to have a global mind-set, and this global perspective should be reflected from the executive level downward to the micro levels of the organization. The next section provides one example of global leadership in practice through the Global Leadership and Organizational Behavior Effectiveness (GLOBE) project.

GLOBE-CLT and Instrumental Leadership for CEOs

In a global study surveying more than 17,000 managers in more than 62 societies, the GLOBE project presented the following results:

- Leaders tend to behave in a manner expected within their country.
- Cultural values do *not* have a direct effect on CEO behavior, rather the effect is indirect through CLTs (culturally endorsed theory—i.e., leadership expectations).
- Both the fit of CEO behaviors (to expectations) and degree of leadership behavior predict effectiveness.
- Superior and inferior CEOs exhibit differing patterns of behavior within their country. (Dorfman et al., 2012: 505)

The composite results from GLOBE's research efforts is a leadership theory called the culturally endorsed theory of leadership (CLT[1]), guided by implicit leadership theory. Implicit leadership theory (Lord, 1977) acknowledges that followers have perceptions or cognitive schemas about what a prototypical leader should entail (Hernandez et al., 2011). In addition, research has highlighted that perceptions of leadership are more related to perceptions of social power than to the actual functions conducted by a leader (Lord, 1977): "the 'structure' in behavior was provided by the cognitive schema of perceivers, not necessarily

---

1   This is different from Complex Leadership Theory (CLT) presented in Chapter 8, Team and Distributed Leadership, in the section "Blended and Hybrid Leadership Model."

the actual behavioral patterns of leaders" (Lord et al., 2016: 121). Other research has found similar results at the group or collective levels. For example, Hogg and Terry (2000) highlighted social processes including in-group and out-group prototypes in which the majority of the perceptions and behaviors were driven by the in-group. This dynamic results in a leadership prototype that is congruent with in-group members and not necessarily with the members of the out-group. This produced the social identity theory of leadership (Hogg, 2001), proposing that "being a prototypical in-group member could be as critical for leadership as being charismatic or having other desirable leadership characteristics" (Hogg, 2001: Self-categorization theory). Implicit leadership theory influences CLT, when dealing with leadership at a global level, it provides a schema in which one's perception of a prototypical leader is highly influenced by the local culture.

The CLT was composed of 6 global dimension and 21 primary dimensions for CEOs. Each of these dimensions are provided in Table 9.3 along with brief descriptions of each global dimension.

The global dimensions of charismatic/value-based, team oriented, self-protective, participative, humane oriented, and autonomous were identified as the main clusters from the data that represented CEOs globally. The charismatic/value-based leadership dimension is similar to other types of leadership: transactional, transformational, and charismatic leadership. This form of leadership is grounded in Vroom's (1964) expectancy theory of motivation, which basically proposed that behavior taken is based on maximizing pleasure and minimizing pain. Performance and motivation depend on an individual's personality, skills, knowledge, experience, and abilities (the traditional KSAs of knowledge, skills, abilities). The global dimension of team oriented aligns with the distributed leadership model presented in TFS along with more specific details relating to teams, teamwork, taskwork, and team effectiveness discussed in Part IV, Team Science. It is important to highlight the fact that new global leadership research efforts, such as the GLOBE project, have found that leadership on a global scale must be a collaborative process rather than an individual one.

The global dimension of self-protective provides safety and security for followers as well as the collective or group. This dimension also includes face saving and status consciousness, which are characteristics of non-Western cultures (Dorfman et al., 2012), incorporating a more international and inclusive list of characteristics.

Participative leadership spans both autocratic and nonparticipative styles of leadership. Both are present around the globe, and leaders must be able to operate in both types of environments, including environments in which both styles are present at the same time and potentially could be a source of conflict. The next global dimension of humane oriented identifies with leaders exhibiting support for their followers along with compassion and generosity (Dorfman et al., 2012). This dimension also could include a leader's level of emotional intelligence or cultural intelligence.

TABLE 9.3.   Global Dimensions

Global Dimensions	Primary Dimensions	Descriptions
I. Charismatic/ Values-based	a. Charismatic 1: visionary	The ability to inspire, motivate, and expect high performance outcomes from others based on core values.
	b. Charismatic 2: Inspirational	
	c. Charismatic 3: self-sacrifice	
	d. Integrity	
	e. Decisive	
	f. Performance oriented	
II. Team oriented	a. Team 1: collaborative team orientation	Emphasized effective team building and implementation of a common purpose or goal among team members.
	b. Team 2: team integrator	
	c. Diplomatic	
	d. Malevolent	
	e. Administratively competent	
III. Self-protective	a. Self-centered	Emphasized safety and security of the individual and group through status enhancement and face saving.
	b. Status conscious	
	c. Conflict inducer (internally competitive)	
	d. Face saver	
	e. Procedural (bureaucratic)	
IV. Participative	a. Autocratic	The degree in which managers involve others in making and implementing decisions.
	b. Nonparticipative	
V. Humane oriented	a. Modesty	Provides supportive and considerate leadership, compassion, and generosity.
	b. Humane oriented	
VI. Autonomous	a. Autonomous	Independent and individualistic leadership attributes (e.g., individualistic, independence, autonomous, uniqueness).

Source: Adapted from Dorfman et al. (2012: 506).

The last dimension, autonomous, relates to a leader's individual characteristics along with one's ability, or preference, to work independently. Although most leaders share these characteristics, albeit at different levels of preferred independence, it is important to realize that autonomy varies around the globe. Although many Western countries value independence and expect leaders to make decisions independently, many non-Western countries value collaborative efforts over independent work. From the perspective of global leadership, it is critical for leaders to realize and value the level of independence or collaboration that is expected of the leader's followers. Similarly, transitioning an organization to a team-based system as well as transforming the leadership ranks to accommodate team-based systems forces some leaders to change their preference from operating independently to being more open and collaborative. This will be a source of resistance during these transformations, as we have highlighted and also will be a source of resistance, for some leaders, when operating in more collective cultures. In either case, leaders need to learn to be more flexible in their leadership styles, which is embedded in the distributed leadership model that is present in TFS.

Conclusion

The chapters included in Part III, Distributed Leadership (Chapters 7, 8, and 9) identified different leadership theories, at different organizational levels, providing options from which organizations can choose. These leadership theories were presented because they work well with one another, from micro levels to macro levels, while also working well with the team-based structures that are all too common in today's working environment. These leadership theories also have been identified as being more effective in dealing with complexity than previous and more traditional leadership theories. It is not intended for any organization to select all of the leadership theories presented in the previous chapters for each organizational level, micro and macro. These chapters present potential leadership theories that could work well with one another from the micro to the macro organizational levels. Having these leadership theories working seamlessly together throughout an organization provides value to the customer and is what we are referring to as distributed leadership for TFS.

Whichever leadership models are found to work well for each level of your organization, it is essential to remember that leadership is a process that fosters emergence; it doesn't command it. This concept is similar to McChrystal et al.'s gardening metaphor:

> I began to view effective leadership in the new environment as more akin to gardening than chess. . . . Within our Task Force, as in a garden, the outcome was less dependent on the initial planting than on consistent maintenance. Watering, weeding, and protecting plants from rabbits and disease are essential for success. The gardener cannot actually "grow" tomatoes, squash, or beans-she can only foster an environment in which the plants do so. (McChrystal et al., 2015: 225)

As in the gardener metaphor, leadership can be fostered by creating the right structures, by enabling safe learning environments, and through training at both the individual and team levels. Leaders can also display "other forms of leadership (e.g., charismatic), [and] enact a mixed profile of leader behaviors" (Antonakis and House, 2014: 749). This should be stressed at each level within an organization, micro and macro, including individuals, teams, external team leaders, and those at the executive levels. Organizations must choose the best leadership theories that work for their contextual setting at each organizational level. These conceptualizations could be simple individual leadership theories or could be more complicated integrated leadership theories. What is essential is that each organization selects the most appropriate model, or integrated model, that works best for them. Ideally, training and development, practice, and time will provide a leadership structure, from the bottom to the top of an organization, that flows bidirectionally. Only then will an organization be closer to achieving flow through a distributed leadership model that works for them.

Reshaping an organizations' ecosystem to better accommodate team-based systems requires changes to the leadership functions of the organization as well as buy-in, and change, at the executive levels of the leadership ranks. This point has been duplicated by others, such as Christensen and Raynor, in describing that changing from existing organizational structures to operate in complex and disruptive environments is essential:

> Corporate executives make this mistake [not changing structures] because most often the very skills that propel an organization to succeed in sustaining circumstances systematically bungle the best ideas for disruptive growth. An organization's capabilities become its disabilities when disruption is afoot. (Christensen and Raynor, 2003: 179)

In realizing that leadership is like a gardener, McChrystal et al. (2015) identified that they began to find success when they started to reshape the ecosystem, facilitated teamwork, and communicated clear strategy and vision for everyone to follow. These steps were found to work when forced to operate in a complex environment (Part II, Complexity Thinking). Reshaping the ecosystem is essentially what this section of the book was about (Part III, Distributed Leadership). The communicating strategy and vision is related to leadership as well. The third component, facilitating teamwork, is what the following section entails (Part IV, Team Science).

References

Antonakis J, Avolio BJ, and Sivasubramaniam N. (2003) Context and leadership: an examination of the nin-factor full-range leadership theory using the Multifactor Leadership Questionnaire. *The Leadership Quarterly* 14: 261–295.

Antonakis J and House RJ. (2014) Instrumental leadership: Measurement and extension of transformational–transactional leadership theory. *The Leadership Quarterly* 25: 746–771.

Bass BM. (1985) *Leadership and performance beyond expectations*. New York, NY: The Free Press.

Bird A and Mendenhall ME. (2016) From cross-cultural management to global leadership: Evolution and adaptation. *Journal of World Business* 51: 115–126.

Boal KB and Hooijberg R. (2000) Strategic leadership research: Moving on. *The Leadership Quarterly* 11: 515–549.

Boal KB and Schultz PL. (2007) Storytelling, time, and evolution: The role of strategic leadership in complex adaptive systems. *The Leadership Quarterly* 18: 411–428.

Christensen CM and Raynor ME. (2003) *The innovator's solution: Creating and sustaining successful growth*, Boston, MA: Harvard Business School Press.

Crossan M, Vera D, and Nanjad L. (2008) Transcendent leadership: Strategic leadership in dynamic environments. *The Leadership Quarterly* 19: 569–581.

Denning S. (2018) *The age of Agile: How smart companies are transforming the way work gets done*. New York, NY: AMACOM.

Dionne SD, Gupta A, Sotak KL, et al. (2014) A 25-year perspective on levels of analysis in leadership research. *The Leadership Quarterly* 25: 6–35.

Dorfman P, Javidan M, Hanges P, et al. (2012) GLOBE: A twenty year journey into the intriguing world of culture and leadership. *Journal of World Business* 47: 504–518.

Dubin R. (1979) Metaphors of leadership: A overview. In: Hunt JG and Larson LL (eds) *Crosscurrents in leadership*. Carbondale: Southern Illinois University Press, 225–238.

Gehrke B and Claes M-T. (2017) Leadership and global understanding In: Marques J and D'himan S-G, Norma (eds) *Leadership today: Practices for personal and professional performance*. Cham, Switzerland: Springer, 371–385.

Gioia DA and Thomas JB. (1996) Identity, image, and issue interpretation: Sensemaking during strategic change in academis. *Administrative Science Quarterly* 41: 370–403.

Hambrick DC and Mason PA. (1984) Upper echelons: The organization as a reflection of its top managers. *Academy of Management Review* 9: 193–206.

Henson R. (2016) *Successful global leadership: Frameworks for cross-cultural managers and organizations*. New York, NY: Palgrave Macmillan.

Hernandez M, Eberly MB, Avolio BJ, et al. (2011) The loci and mechanisms of leadership: Exploring a more comprehensive view of leadership theory. *The Leadership Quarterly* 22: 1165–1185.

Hogg MA. (2001) A social identity theory of leadership. *Personality and Social Psychology Review* 5: 184–200.

Hogg MA and Terry DJ. (2000) Social identity and self–categorization processes in organizational contexts. *Academy of Management Review* 25: 121–140.

Lord RG. (1977) Functional leadership behavior: Measurement and relation to social power and leadership perceptions. *Admiistrative Science Quarterly* 22: 114–133.

Lord RG, Gatti P, and Chui SLM. (2016) Social-cognitive, relational, and identity-based approaches to leadership. *Organizational Behavior and Human Decision Processes* 136: 119–134.

Lovelace JB, Neely BH, Allen JB, et al. (2019) Charismatic, ideological, & pragmatic (CIP) model of leadership: A critical review and agenda for future research. *The Leadership Quarterly* 30: 96–110.

McChrystal S, Collins T, Silverman D, et al. (2015) *Team of teams: New rules of engagement for a complex world*. New York, NY: Penguin.

Morgeson FP, DeRue SD, and Karam EP. (2010) Leadership in teams: A functional approach to understanding leadership structures and processes. *Journal of Management* 36: 5–39.

Nossel M. (2018) *Powered by storytelling: Excavate, craft, and present stories to transform business communication.* New York, NY: McGraw-Hill.

Steers RM, Sanchez-Runde C, and Nardon L. (2012) Leadership in a global context: New directions in research and theory development. *Journal of World Business* 47: 479–482.

Uhl-Bien M and Marion R. (2009) Complexity leadership in bureaucratic forms of organizing: A meso model. *The Leadership Quarterly* 20: 631–650.

Vroom VH. (1964) *Work and motivation.* New York, NY: Wiley.

Waldman DA, Javidan M, and Varella P. (2004) Charismatic leadership at the strategic level: A new approach of upper echelons theory. *The Leadership Quarterly* 15: 355–380.

Waldman DA and Yammarino FJ. (1999) CEO charismatic leadership: Levels-of-management and level-of-analysis effects. *Academy of Management Review* 24: 266–285.

PART IV

Team Science

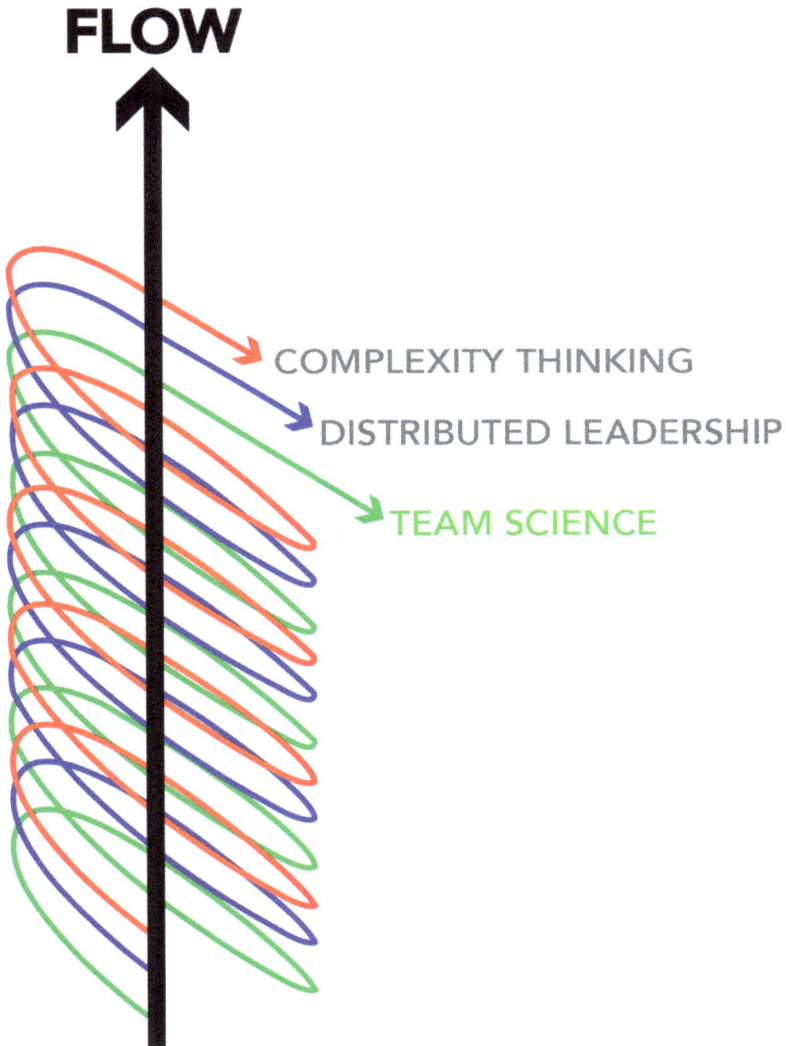

FLOW

COMPLEXITY THINKING

DISTRIBUTED LEADERSHIP

TEAM SCIENCE

CHAPTER 10

Teams, Teamwork, and Taskwork

Team Science

Team science is a multidisciplinary field that concentrates on the interpersonal, intrapersonal, organizational, physical environmental, technological, societal, and political contextual factors in the workplace. Team science touches on the collaborative functioning of teams and small groups in the workplace, often involving cross-disciplinary and cross-functional groups. The field of team science is a multidisciplinary field of study in that it is supported by research conducted in a variety of disciplines (e.g., ecology, healthcare, organizational science, psychology) and covers multiple contextual settings. First, *team science* (TS) combines empirical findings from multiple disciplines to address important issues (e.g., health and organizational problems, complex, and wicked problems) that require the use of teams to derive solutions to current problems. The TS field is associated mostly with team research in organizational settings. In addition, the empirical area of inquiry known as the *science of team science* (SciTS) concentrates on collaborative efforts to better address "complex environmental, social, and health problems" (Falk-Krzesinski et al., 2011: 145). Through the efforts of the National Institute of Health (NIH) and the National Cancer Institute (NCI), the SciTS was formed as an effort to understand and manage "circumstances that facilitate or hinder the effectiveness of large-scale, cross-disciplinary, collaborative research initiatives" (Falk-Krzensinski, 2012: The science). The SciTS is most associated with research teams in the sciences and healthcare settings.

The TS field has evolved to identify evidence-based methods to better conduct collaborative research using newly developed problem-solving methodologies and techniques. Team science is essentially a "cross-disciplinary science in which team members with training and expertise in different fields work together to combine or integrate their perspectives in a single research endeavor" (National Cancer Institute, n.d.: About team science). TS, as it relates to the workplace, focuses on how to better deal with complex problems and shared resources while operating among interdisciplinary, cross-cultural, and cross-functional teams in a global environment. Utilizing the empirical knowledge gained from both fields of study, TS and SciTS, The Flow System (TFS) applies this knowledge for organizations and institutions to maximize the benefits provided by using team-based structures in addressing complexity and disruptive environments. Team science, within TFS, concentrates on understanding and enhancing the antecedent conditions (inputs), collaborative processes (teamwork), and outcomes (team effectiveness; Stokols et al., 2008) associated with team-based structures in organizational and institutional settings.

From this body of research, the TS component of TFS will be presented in Part IV, Team Science. The next section of this chapter will begin with a discussion of teams in contrast to groups, will explain why teams are necessary, and will provide a discussion of the two components of teams: teamwork and taskwork.

Teams

Teams are becoming more prevalent in the workplace as well as in the research literature. In the workplace, teams have emerged as the organizational building blocks of today's businesses in which failing to "value and invest in teamwork can have catastrophic consequences" (Salas et al., 2017: 4). In the literature, "science is paying attention to teams" (Fiore, 2008: 253) with a renewed interest on collaborative processes. Research on teams has grown in recent decades (Falk-Krzesinski et al., 2011) with a renewed interest in collaborative tools to address complex phenomenon (Fiore, 2008). Findings have been consistent in that teamwork is becoming a more essential component for success in organizational and research settings, spanning multiple disciplines, including "science and engineering, social sciences, the arts and humanities, and patents" (Fiore, 2008: 258).

But what constitutes a team, and when and why are teams needed? In general, a basic definition for a team could be that teams are composed of more than one individual working on a common goal. A more formal definition of a team from Cohen and Bailey follows:

> A team is a collection of individuals who are interdependent in their tasks, who share responsibility for outcomes, who see themselves and who are seen by others as an intact social entity embedded in one or more larger social systems (for example, business unit or the corporation), and who manage their relationships across organizational boundaries. (Cohen and Bailey, 1997: 241)

Kozlowski and Ilgen provide a second definition of a team, one that includes specific components for a team, as follows:

A team can be defined as (a) two or more individuals who (b) socially interact (face-to-face or, increasingly, virtually); (c) possess one or more common goals; (d) are brought together to perform organizationally relevant tasks; (e) exhibit interdependencies with respect to workflow, goals, and outcomes; (f) have different roles and responsibilities; and (g) are together embedded in an encompassing organizational system, with boundaries and linkages to the broader system context and task environment. (Kozlowski and Ilgen, 2006: 79)

In looking at both definitions, it is clear that a team is more than a few people working on the same or similar tasks. A team requires shared goals with team members working independently[1] as well as interdependently[2] toward a common, overarching goal. Contrast this with groups, which are often mistaken for teams in organizations today. A group involves a number of people, typically more than two, working on the same or similar tasks. Working groups are composed of "people who learn from one another, share ideas, but are not interdependent in an important fashion and are not working toward a shared goal" (Thompson, 2018: 5). Some working groups, by definition, could take the form of communities of interest, communities of practice, or coaching groups (Hackman, 2011). Although each of these specific groups have more than two people who have joined together because of a common interest, they fail to meet the definition of a team because group members do not have a shared goal requiring each member to share information, to work independently and interdependently, to coordinate activities, and to share roles and responsibilities, resulting in achieving a shared goal of the group.

Teams are composed of individuals working on interdependent tasks that contribute to the overall task objective of the collective. The first identifier of a team is the interdependency of individual members working toward a common goal. Second, team members are required to interact to combine each team member's interdependent portion of the overall task. To combine each individual's effort into a composite that represents the team's output, team members must interact with one another. This interaction involves shared responsibilities in which authentic communication (Kayser, 2011) is a requirement. Last, a team also must be adaptive (Cohen and Bailey, 1997). Today's knowledge economy presents complex dynamic problems to which teams need to adapt. Although adaptability may not be an initial requirement for a team, if a team is to remain successful in "a dynamic, shifting, and complex environment" (Kozlowski and Ilgen, 2006: 78), it must learn to be adaptive.

---

1   Independent refers to a team member not depending on any other team member to complete a task or assignment.

2   Interdependence refers to "the extent to which a job is contingent on others' work and other jobs are dependent on the work of the focal job" (Cordery and Tian, 2017: 111).

A team that cannot be adaptive because too many restrictions are placed onto the team—for example, by being bounded with top-down, command-and-control directions from management—becomes inhibited, dampening a team's ability to experiment and innovate (Forsgren et al., 2018). Teams need some level of autonomy, allowing them to experiment and try new ideas, to take risks, to self-organize and reach a level of emergence that is necessary in complex environments. In cases in which teams are stifled because of too many restrictions and procedures, preventing them from functioning as an autonomous entity, then using teams as the organizational structure may not be the best course of action. In reality, in environments in which teams cannot function as autonomous units, the organization is operating using groups and calling them teams. This practice can be found throughout corporations today around the globe, in which groups are operating as teams by name only (Denning, 2018), resulting in groups operating inconsistently with suboptimal performance. This problem, of groups operating as teams, could result in mediocracy: "Mediocre teams and companies don't know what they are. They say one thing and then do another. They blow around like the wind, and they destroy authenticity and trust" (Karlgaard and Malone, 2015: 93).

Utilizing shared leadership as the team's leadership structure, a team should be composed of the following five basic functions, at a minimum:

1. Compelling performance purpose. Exceeds sum of individual goals.
2. Members work jointly to integrate complementary talents and skills.
3. Work products (outcomes) are mostly collective or joint efforts.
4. Adaptable working approach shaped and enforced by members.
5. Mutual plus individual accountability. (Katzenbach and Smith, 2001: 4)

Team performance and learning is emergent, and it cannot be represented as the sum of each individual team member. A team's performance and its ability to learn involves some degree of shared-ness, shared cognition, and learning transfer, which makes these constructs greater than the sum of the team's individual parts, in this case, the sum of individual team members. This shared-ness comes from team members interacting and combining skills and resources to achieve a team's goals. A team's outcome is a collective product that is accomplished as a team overcomes obstacles, allowing team members to adapt and self-organize accordingly. Team members are also accountable for their own actions and also are accountable for one another. These five functions present the basic components that make a team a team, as opposed to a group. More detailed information on what makes a team will be presented in the following sections of this chapter. Teams are composed of two essential components, teamwork and taskwork, each with their own components and purposes. Team effectiveness is a composite of teamwork, taskwork, the team's performance, and the value delivered to the customer, making team effectiveness a more comprehensive indicator of a team's success. Team effectiveness will be covered in Chapter 11, Team Effectiveness.

But first, the next section discusses why teams are beneficial and how to determine the best size for a team in today's complex environment.

Why Teams

Within the context of TFS, operating in complex environments, we want to introduce a third definition of a team: "Complex dynamic systems that exist in a context, develop as members interact over time, and evolve and adapt as situational demands unfold" (Cohen and Bailey, 1997: 78). This definition positions teams as complex adaptive systems that are self-organizing and adaptive, leading to emergence (evolving) if the situation demands a new functional unit or structure.

Leaders and managers in organizations need to be cognizant of when to utilize a team compared with assigning individual tasks. Managers must be able to balance team assignments, and they must avoid assigning everything to teams (groups) (Hackman, 2011) while also trying not to avoid teams all together. Teams can become ineffective if managed poorly or if they become too large (e.g., groupthink, social loafing). Teams are a more logical choice than individual assignments for specific tasks or problems. Again, similar to selecting the right tool for the job, leaders must identify when to use teams and when not to use teams, choosing the right structure for the job.

Why Do We Need a Team?

The question that should be asked, first, is *Why do we actually need a team?* (Hackman, 2011). Through experience and research on teams, Hackman provided the following questions to ask when determining whether a team is needed:

> Is it because the task requires more resources than any one person can provide? Or because diverse skills and perspectives are required to accomplish the work? Or because flexibility is needed to keep pace with a rapidly changing context? Or because you want to provide a setting in which individual members can hone their personal capabilities through interactions with others? If none of these reasons applies, there probably is no need for the extra work and leadership attention that it takes to create and support a team. (Hackman, 2011: 27)

Teams are capable of assigning, and reassigning, tasks to react to environmental changes, are more diverse in knowledge and skills, have access to more resources, and are better able to address complex problems. Some tasks are simple or repetitive and are categorized as being at the lower levels of the complicated domain (Cynefin framework). These tasks do not necessarily require whole teams.

We would add to Hackman's list of questions the contextual setting: What type of problems are being expected? and What type of environment will be experienced? If complex or wicked problems are expected, or if the environment is expected to potentially be a complex environment, then we propose that teams are the best organizational structure that could be assembled. Utilizing a diverse set of expertise (knowledge, skills, and experiences) among a small number of people, as in a team setting, has been shown to be a prerequisite for solving complex problems (Slyngstad et al., 2017). Here, small and diverse teams are required when addressing complex environments or problems. In many cases, when teams fail, they fail for two main reasons: lack of support and resources from management and the team is too large. This second point was highlighted by Karlgaard and Malone in their research: "When teams fail, they are almost always too big" (Karlgaard and Malone, 2015: xix).

Regarding the foundation of TFS, providing value to the customer, teams are better able to communicate directly with the customer because of their small size, flexibility, and entrepreneurial spirit (Rigby et al., 2018). Teams are closer to the customer and communicate directly with the customer and are the first to receive feedback, providing them with the opportunity to revise a product to better meet the customer's needs quickly. This process, at times, can be expedited when the customer becomes a team member. This provides a direct connection with the customer throughout the team's tenure.

Given that teams are today's primary structure selected by organizations, institutions, and government agencies and that team working skills have become one of the essential skills required of today's leaders, it must be acknowledged that teams have become an effective organizational unit, when done correctly. Conducting teamwork correctly can become an organizations' competitive advantage: "If a company wants to outstrip its competitors, it needs to influence not only how people work but also how they work together" (Duhigg, 2016: Our data-saturated).

Team Size

Responding to today's complex problems, organizations are using virtual tools and technologies to take advantage of employees' specialized skills and to enhance their contributions to organizational teams (Wu et al., 2019). A research study that reviewed more than 65 million papers, patents, and software products reviewed the level of disruption from these collaborative ventures. Findings from this study showed that true disruptive research[3] came from smaller teams rather than larger teams: "These results support the hypothesis that large teams may be better designed or incentivized to develop current science and technology, and that small teams disrupt science and technology with new problems and opportunities" (Wu et al., 2019: 379). As the size of the research team increased, the level of disruption that was found had diminished significantly (Wu et al., 2019).

---

3   This study contrasted developing papers with disruptive papers. Developing referred to papers that included "extensions or improvements of previous theory, method or findings," whereas disruptive referred to those that provide "punctuated advances beyond previous theory, methods or findings" (Wu et al., 2019: 384).

These findings show that smaller teams are more capable of challenging existing practices and policies. Smaller teams are less likely to conform and are more likely to interact and have fruitful exchanges around current practices as opposed to taking practices for granted. Smaller teams are less likely to favor one discipline's methods and techniques over another because no one discipline is overrepresented. Smaller teams hold the potential of "expanding the frontiers of knowledge, even as large teams rapidly develop them" (Wu et al., 2019: 382).

When addressing complex problems, one technique that has been found to be successful is to produce prototypes[4] (Elverum and Welo, 2016; Lord et al., 2016). Prototypes are models that can be tested, providing evidence of the prototypes strengths and flaws. Lessons from one prototype can be applied to the next prototype, providing a series of design and testing stages to model the complex problem. When this process is carried out in a continuous manner, it is called rapid prototyping. Toyota gained a competitive advantage through their capabilities of being capable of producing more prototypes, resulting in improvements in development time and product quality (Elverum and Welo, 2016). Rapid prototyping is commonplace in new product development teams in which each team member has specialized skills that concentrate on the prototype's characteristics. Rather than concentrating on the size of a team, what should be the determining factor is to identify the required competencies and skills to complete the team's tasks. Build a team, not on a predetermined number or size, but on the requisite skills required for the task or prototype.

Team Transition Processes

Before becoming an effective team member and before one can lead or manage a team, one must first understand the processes that take place during a team's tenure, from formation to dismantling the team. Team processes consist of recurring processes, meaning that similar processes occur for each effective team. These processes have been articulated into a model of team transition processes from Marks et al. (2001). These transition processes represent the concept that team processes are "multidimensional and that teams use different processes simultaneously and over performance episodes in order to multitask effectively" (Marks et al., 2001: 362). This model breaks down the essential team processes into three phases: transition phase processes, action phase processes, and interpersonal processes (Marks et al., 2001). These phases are illustrated in Figure 10.1 with the components for each phase provided.

---

4 A prototype can be defined as "a concrete representation of part or all of an interactive system" (Elverum and Welo, 2016: 3007). Prototypes can consist of analytic or physical models, including "sketches, mathematical models, simulations, test components and fully functional preproduction versions of the product" (Elverum and Welo, 2016: 3007).

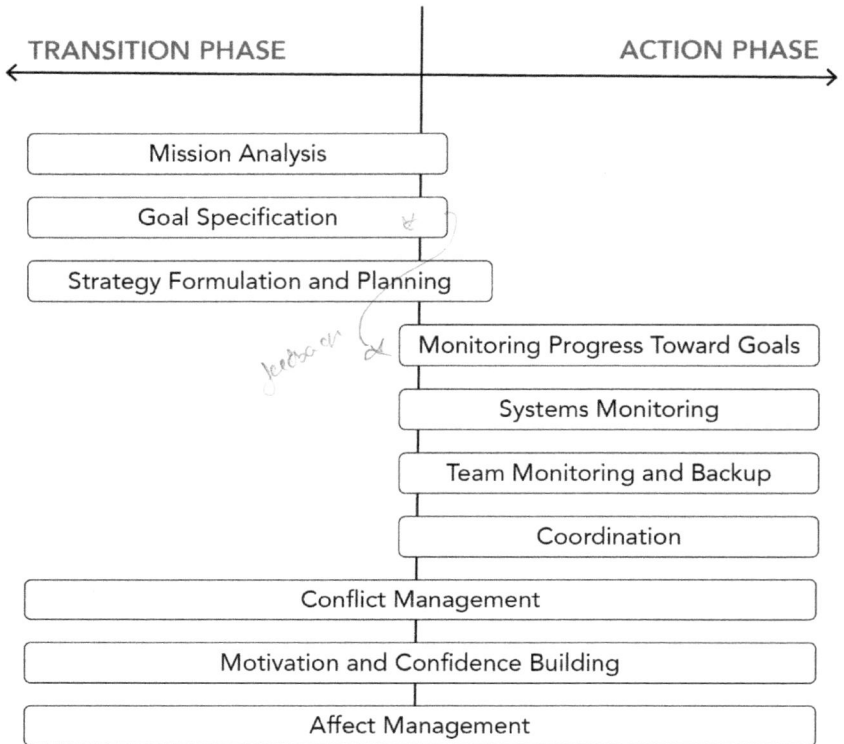

FIGURE 10.1.   Team Transition Processes

Transition Phase Processes

The transition phase processes include those processes that involve evaluation or planning of activities toward goal attainment. The processes that make up the transition phases include mission analysis, goal specification, and strategy formulation and planning. The strategy formulation and planning processes also includes three subdimensions: deliberate planning, contingency planning, and reactive strategy adjustment and transmission of a principal course of action for mission accomplishment (Marks et al., 2001). These processes are defined in Table 10.1.

Action Phase Processes

The action phase processes include those processes that are directly related to goal accomplishment. These processes include monitoring progress toward goals, systems monitoring, team monitoring and backup response, and coordination activities (Marks et al., 2001). These processes are defined in Table 10.1.

TABLE 10.1. Team Transition Phases

Transition Processes	Description
Transition Phase Processes	Evaluating and planning activities aimed at achieving the team's goals.
Mission Analysis	Interpretation and evaluation to the team's mission and tasks.
Goal Specification	Identification and prioritization of goals.
Strategy Formulation and Planning	Developing course of action for accomplishing the team's mission.
Deliberate Planning	Formulating action toward goal accomplishment.
Contingent Planning	Discussion of alternative plans and strategies during initial planning stages.
Reactive Strategy Adjustment	Reactions to unanticipated changes.
Action Phase Processes	Activities toward achieving a team's goals.
Monitoring Progress Toward Goals	Tracking and interpreting progress.
Systems Monitoring	Tracking team resources and environmental conditions.
Internal Systems Monitoring	Tracking personnel, equipment, and other information generated by the team.
Environmental Monitoring	Tracking external conditions.
Team Monitoring and Backup Responses	Assisting team members perform tasks through feedback, assisting one another, and completing assigned tasks.
Coordination Activities	Orchestrating the sequence and timing of interdependent actions.
Interpersonal Processes	Processes that govern interpersonal activities.
Conflict Management	Avoiding dis-functional conflict.
Preemptive Conflict Management	Establishing conditions to prevent, control, or guide team conflict before it occurs.
Reactive Conflict Management	Working through task, process, and interpersonal disagreements.
Motivating/Confidence Building	Generating and preserving a sense of collective confidence.
Affect Management	Regulating emotions (e.g., social cohesion, frustration, excitement).

Source: Marks et al. (2001).

Interpersonal Processes

The interpersonal processes often emerge at the beginning of a team's formation and evolve throughout that team's tenure until the team is dismantled. These interpersonal processes, when ignored, can lead to a team being dysfunctional, as most interpersonal processes must be addressed and managed at the beginning of a team's formation. The interpersonal processes include conflict management, motivation and confidence building, and affect management, which all represent "processes teams use to manage interpersonal relationships" (Marks et al., 2001: 368). These processes are also defined in Table 10.1.

The team transition processes incorporate both teamwork and taskwork activities involved in one taxonomy, showing that teamwork and taskwork activities are interconnected and must be monitored and managed for successful team outcomes to occur. This model also introduces time as an important factor in the team processes and highlights how different processes play out during different times of a team's tenure. Embedded within this temporal model are the essential components of any team: team composition, taskwork, and teamwork. The following sections provide additional details on each of these components.

Team Composition

Team composition is essential in ensuring that the right mix of team member diversity is achieved to meet the demands of the projected team tasks. Here, team diversity is described as team member knowledge, skills, and abilities required for a team to be successful (Wolfson and Mathieu, 2017). In contrast, team composition is defined as the configuration of team member knowledge, skills, abilities, and other characteristics (KSAO) (Wolfson and Mathieu, 2017). Team composition has positive influence on team processes and outcomes. To ensure that a team is capable of transitioning through each of the team transition processes highlighted by Marks et al. (2001), a team must be composed with the requisite knowledge, skills, and abilities along with any other characteristic that the team might need for goal attainment.

At various points in a team's tenure, specific team members could be considered its core member with other team members being viewed as being noncore members. Core team members could be perceived by other team members because of the core member's skills or experiences. Core members may be conceived as superstars, whereas noncore members are merely in the background. Teams composed of too many superstars, or teams that have too many members striving for core roles, typically are not successful (Bolinger et al., 2018). Team membership must consist of members who are capable of taking on the role of both a core member and a noncore member in which noncore members support the core member's role, resulting in an "interdependent mix of members in strategically core and noncore roles who work together effectively" (Bolinger et al., 2018: 680).

Team composition in disruptive and complex environments is most effective when utilizing a *team profile model*. A team profile model adopts a collective perspective, with the primary focus at the team level, in which the KSAOs of the team are considered rather than focusing on the KSAOs of the individuals (Wolfson and Mathieu, 2017). Among the other characteristics in the KSAOs, research has concentrated on a member's network of interactions and communication (Wolfson and Mathieu, 2017), a team member's reach, and a team's external influence. At times, information and external resources will be required, and a team with some external influence will have a competitive advantage over a team without external influence. Teams should not be composed of political players but rather should be composed of some members who have connections or networks external to the team.

Team composition must consider the tasks that team members will be performing, the teamwork and interpersonal relationships expected among the team members, and the ability of team members to obtain external resources when needed. Team composition has the potential to lead to successful team outcomes: "The extent to which individuals within a team have greater task specific, and generic teamwork oriented capabilities, should lead to greater effectiveness in team functioning and performance" (Marks et al., 2001: 131). When composing any team, the taskwork and teamwork functions that will be expected of the team members must be considered. All requisite KSAOs must be available along with team members who have the ability to interact and network outside of the team, when needed. The components of taskwork and teamwork will be discussed next.

Taskwork

Taskwork is defined as "work done by an individual, or a group of individuals, behaving autonomously rather than interdependently" (McIntyre and Salas, 1995: 33). It also has been referred to as "the ability to perform individual tasks" (Kozlowski and Ilgen, 2006: 95). Perhaps, one of the most common definitions of taskwork describes "functions that individuals must perform to accomplish the team's task" (Mathieu et al., 2008: 420) or to achieve a team's goals (Dihn and Salas, 2017). Tasks also include those organizational needs that are accomplished using team-based structures (Wildman et al., 2012).

Taskwork involves the technical aspect of what work gets done, involving the processes and procedures required to complete a specific task. Although taskwork is associated with task-based functions, teamwork deals more with interpersonal relationships between team members (see the section "Teamwork"). Taskwork requires knowledge of team members to understand the required processes used to complete the task and the procedures involved, sometimes requiring only one individual while other times requiring more than one individual. Taskwork involves team members communicating and coordinating activities with each other. Team level activities include the following:

> *Orientation* (i.e., information acquisition and exchange), *resource distribution* (i.e., allocation of members and resources to roles, workload balancing), *timing* (i.e., activity pacing), *response coordination* (i.e., sequencing and synchronization of member actions), *motivation* (i.e., goal setting, incentives, and conflict resolution), *systems monitoring* (i.e., error checking), and *procedures maintenance* (i.e., developing and maintaining performance standards). (Devine, 2002: 306)

Teams utilize both individual taskwork as well as team-based taskwork activities, and the type of tasks can vary depending on the type of team. A variety of different team types are found in the literature: project, advice, parallel, problem-solving, suggestion, cross-functional, improvement, small-project, special-purpose, or task-force teams (Devine, 2002). The information provided in this section focuses primarily on teams found in the workplace: intellectual work teams, command teams, negotiation teams, commissions, design teams, advisory teams, physical work teams, production teams, performance teams, medical teams, response teams, military teams, and transportation groups (Devine, 2002).

Tasks associated with teams in the workplace have been categorized as integrated tasks that can be completed by individual team members or by the team as a whole, and by higher-level attributes that only take place at the team level. Table 10.2 highlights each of these task types.

TABLE 10.2. Task Types

Task Type	Description
Integrated tasks	Different types of work tasks that teams (or individuals) can engage.
Managing others	Directing, supervising, or overseeing the work of others in an authoritative role.
Advising others	Providing professional support, such as expert assistance or advice, in a consultative role in which the advisor lacks authority over those who he or she is advising.
Human service	Social interaction where an individual or team is providing a good or service to another party.
Negotiation	Social interaction in which two or more parties in conflict seek to resolve differences and reach agreement.
Psychomotor action	Technical or motor functioning requiring psychological processing to perform calculated or elaborate movements, including the manipulation, operation, or use of a product, machine, or object, or a task that is achieved by engaging in psychomotor action of some sort.
Defined problem solving	Problem-solving tasks with predetermined or conclusive solutions or correct answers.
Ill-defined problem solving	Problem-solving tasks lacking predetermined or conclusive solutions or correct answers, such as idea, plan, or knowledge generation.
Team-level tasks	Holistic attributes that describe the team as a whole.
Task interdependence	The extent to which outcomes of the team members are influenced by, or depend on, the actions of others.
Role structure	The extent to which roles are fundamentally different and therefore not interchangeable or each person is capable of performing every component.
Leadership structure	The pattern, or distribution, of leadership functions, such as setting direction and aligning goals, among the members of the team.
Communication structure	The pattern, or flow, of communication and information sharing among the members of the team.
Physical distribution	The spatial location of the team members in reference to one another.
Team life span	The length of time for which the team exists as a functional, active unit.

Source: Wildman et al. (2012).

As teams are formed for a variety of reasons and to complete a number of different tasks, the technical aspects of tasks expected of team members (e.g., programming, engineering, surgery) vary drastically. Tasks are contextual to each setting and team. Beyond the technical aspects of tasks expected of teams, the integrated tasks and the team-level tasks presented in Table 10.1 provide tasks that must be considered and managed while also conducting the contextual technical aspects of a team's tasks. Concentrating solely on the technical tasks provides little insight into the core team characteristics required to complete a team's tasks successfully, "illustrating the importance of considering core team characteristics in conjunction with task type" (Wildman et al., 2012: 120). The task type along with the requisite team characteristics necessary to complete the team's task also should be considered in the team composition stage, ensuring that the KSAO and other characteristics of the team members are suitable to completing the team's tasks.

Tasks also differ because of their level of complexity or lack of complexity. For simple or complicated tasks, teams composed functionally[5] are more effective. For simple and complicated (no complexity or very low levels of complexity) problems, teams operate using either a pooled task structure in which team members work in parallel or in a sequential task structure in which tasks are addressed sequentially (one at a time) (Kozlowski et al., 2015). These routine tasks (simple) require limited levels of experience and knowledge and produce less effort and resources (Mohammad et al., 2015). Both types of task structures, pooled and sequential, work best for stable and predictable task environments due to the weak linkages and couplings required of team members (Kozlowski et al., 2015).

For dynamic and complex tasks, teams composed divisionally[6] are more superior (Kozlowski et al., 2015). For complex problems, reciprocal task structures are more adaptable and flexible while providing feedback and support for team members to make adjustments when needed (Kozlowski et al., 2015). As the level of complexity increases, the effort, skill level (KSA), and resources also increase, requiring more distribution divisionally to address problems using a reciprocal task structure.

Alignment among a team's task type (simple, complicated, complex) represents the team's external task structure. This external task structure must align with the team's internal task structure and with the team members' individual

---

5 Functionally represents team members with limited and highly specialized skills that complement each other (Kozlowski et al., in press). Functional structures require a high level of coordination and mutual support (Hollenbeck et al., 2002).

6 Divisional composition identifies team members with broad skills and takes multifaceted roles, enabling members to work independently as well as interdependently (Kozlowski et al., in press). Divisional structures place a premium on cognitive ability (Hollenbeck et al., 2002).

differences in KSAs. Team composition and task structure must be in alignment, thus providing "a good external fit between the team and the task environment" (Hollenbeck et al., 2002: 605).

Beyond team composition and the team's expected taskwork, a team must be capable of conducting work as a cohesive and holistic unit. This takes us to the most important aspect in determining whether or not a team will be successful—that is, teamwork. We turn to this discussion in the next section.

Teamwork

Teamwork is a dynamic process that changes and develops over time (McIntyre and Salas, 1995). Some of the more common features of teamwork that should improve over time, through team member's experiences, practice, and training include "performance, monitoring, feedback, closed-loop communication, backing-up behavior, team awareness, and within-team interdependence" (McIntyre and Salas, 1995: 32).

Teamwork is a composite of "shared behaviors, attitudes, and cognitions" (Dihn and Salas, 2017: 16) that allow a team to achieve its goal. By this definition of teamwork, three psychological foundations are involved in teamwork:

- attitude,
- behaviors, and
- cognitions (Dihn and Salas, 2017).

Attitudes relate to the internal conditions that can interfere with team member interactions, while behaviors are associated with the processes required to conduct teamwork. Cognition relates to team members shared knowledge and the structure of that knowledge (Dihn and Salas, 2017).

As a team construct, teamwork is a product of all of the team members and not a product of one, or a few, team members. Interactions between team members is an essential component of effective teamwork; teams will be incapable of accomplishing their goals if interactions do not take place among team members. Team members must interact, "must work toward valued, common, specified goals and objectives; and must adapt to circumstances in order to meet these goals and objectives" (McIntyre and Salas, 1995: 23). The following principles were highlighted by McIntyre and Salas to show the different requirements of teamwork:

> Principle 1: *Teamwork means that members monitor one another's performance.*
> Principle 2: *Teamwork implies that members provide feedback to and accept it from one another.*
> Principle 3: *Teamwork involves effective communication among members, which often involves closed-loop communication.*
> Principle 4: *Teamwork implies the willingness, preparedness, and proclivity to back fellow members up during operations.*

Principle 5: *Teamwork involves group members' collectively viewing of themselves as a group whose success depends on their interaction.*
Principle 6: *Teamwork means fostering within-team interdependence.*
Principle 7: *Teamwork is characterized by a flexible repertoire of behavioral skills that vary as a function of circumstances.*
Principle 8: *Teams change over time.*
Principle 9: *Teamwork and taskwork are distinct.* (McIntyre and Salas, 1995: 23–33)

New research is providing renewed support for teamwork over taskwork (Kozlowski and Ilgen, 2006). This research is counterintuitive to industry thinking. One view believes that a team composed of individuals capable of completing the required tasks will be most successful. A contrasting perspective believes that teams composed of members capable of performing teamwork skills will perform better than teams composed solely of members capable of performing tasks. Team members' ability to perform most of the tasks required of the team is essential; however, it is becoming more essential that teams be able to perform basic teamwork functions. Without basic teamwork skills, a team will be less successful, regardless of their ability to perform the individual tasks expected of the team. Teamwork initiates interaction and the free flow of information, providing shared resources and knowledge that can be applied to any one task, thus providing a team with more ability when compared with each individual team member's skills. This aligns with the concept of emergence in which the total capabilities of the team are greater than the individual team members' capabilities—and teamwork is the secret sauce to effective teams. Other researchers have come to the same conclusion: "This begs the question, what ensures the success of a team? We submit the answer is teamwork" (Salas et al., 2005: 556).

The Dimensions of Teamwork

The dimensions of teamwork are composed of two distinct factors:

- core processes and emergent states, and
- influencing conditions (Dihn and Salas, 2017).

Core processes and emergent states involve essential processes required to convert inputs into team outcomes, involving affective, behavioral, and cognitive mechanisms (Dihn and Salas, 2017). This also can be referred to as internal team processes. These core processes and emergent states include cooperation, conflict, coordination, communication, coaching, and cognition. Influencing conditions involve external forces that influence the teamwork processes, involving composition, context, and culture (Dihn and Salas, 2017).

Collectively, these dimensions of teamwork have been identified as the nine Cs of teamwork. Table 10.3 lists the six core processes and emergent state along with a brief description of each. Table 10.3 also provides additional subdimensions (i.e., related constructs) that are relevant to each dimension.

TABLE 10.3.   Teamwork Core Processes and Emergent Factors

Core Processes and Emergent Factors	Description
Cooperation	The motivational drivers of teamwork. In essence, this is the attitudes, beliefs, and feelings of the team that drive behavioral action (Dihn and Salas, 2017: 17). Motivation and desire to engage in coordinative and adaptive behavior (Weaver et al., 2013).
Goal Commitment	The determination to achieve team goals (Dihn and Salas, 2017: 19).
Checking In	A discussion of prior and relevant experiences before task performance (Dihn and Salas, 2017: 19).
Psychological Safety	The shared feeling of safety within a team allowing for interpersonal risk taking (Dihn and Salas, 2017: 19). A shared belief in which the team is safe from risk taking (Edmondson, 1999). When colleagues trust and respect each other and feel able—even obligated—to be candid (Edmondson, 2019: 8).
Trust	The willingness of a party to be vulnerable to the actions of another of another party based on the expectation that the other will perform a particular action important to the truster, irrespective of the ability to monitor or control that other party (Mayer et al., 1995: 712).
Cohesion	Social cohesion, how members are attracted to one another, and task cohesion, the shared commitment among members to achieve a goal (Forsyth, 2014: 137).
Conflict	The perceived incompatibility in interests, beliefs, or views held by one or more team members (Dihn and Salas, 2017: 23; see also Jehn and Shah, 1997). Differences or incompatibilities in interests, values, power, perception, and goals (Yasmi et al., 2006).
Cognition Conflict	Team member cognitive states (overlapping cognitive representation of team member knowledge, team member representation of tasks, equipment, working relationships, and situations) (Turner, 2016: 161).
Process Conflict	How well groups manage and coordinate activities (Behfar et al., 2010)
Relationship Conflict	Affective, emotional, and focused on personal incompatibilities or disputes (Amason, 1996).
Status Conflict	Disputes over the relative status positions in a team's social hierarchy (Greer and Dannals, 2017: 327).
Task Conflict	Describes disagreement about the work that is being done in the group (Jehn and Chatman, 2000).
Coordination	The process of orchestrating the sequence and timing of interdependent actions (Marks et al., 2001).
Explicit	Team member directly and intentionally plan and communicate to manage interdependencies (Dihn and Salas, 2017; Rico et al., 2008).

TABLE 10.3. Teamwork Core Processes and Emergent Factors (Continued)

Core Processes and Emergent Factors	Description
Implicit	Team members communicating to articulate plans, define responsibilities, negotiate deadlines, and seek information (Rico et al., 2008: 165).
Communication	A reciprocal process of team members' sending and receiving information that forms and reforms a team's attitudes, behaviors, and cognitions (Dihn and Salas, 2017: 17).
Coaching	The enactment of leadership behaviors to establish goals and set direction that leads to the successful accomplishment of these goals (Dihn and Salas, 2017: 17).
	Helping individual members learn ways they can strengthen their personal contributions and, at the same time, exploring ways the team as a whole can make the best possible use of its member resources (Hackman, 2011: 135).
Cognition	A shared understanding among team members that is developed as a result of team member interactions, including knowledge of roles and responsibilities; team mission objectives and norms; and familiarity with teammate knowledge, skills, and abilities (Dihn and Salas, 2017: 17).
Shared Cognition	Team members' shared understanding of team tasks, equipment, roles, goals, and abilities (Lim and Klein, 2006: 403).
Transactive Memory System	Memory that is influenced by knowledge about the memory system of another person (Lewis, 2003: 588).
	A collection of knowledge possessed by each team member and a collective awareness of who knows what (Mathieu et al., 2008: 431).
Shared Mental Memory	A knowledge-based team competency essential to team effectiveness (Dihn and Salas, 2017: 24).
	An organized understanding or mental representation of knowledge that is shared by team members (Mathieu et al., 2005: 38).
Task-Related Features	Task-related features of the situation (Mathieu et al., 2005).
Team-Related Aspects	Team aspects of the situation that include team interaction and team models (Mathieu et al., 2005).

Google conducted an internal study called *Aristotle* to look at what leads to successful teams. In their findings, Google reported the following: "The paradox, of course, is that Google's intense data collection and number crunching have led it to the same conclusions that good managers have always known. In the best teams, members listen to one another and show sensitivity to feelings and needs" (Duhigg, 2016: The technology industry).

Although this statement could apply to any of the core processes and emergent factors identified in Table 10.3, it especially speaks to the fact that teamwork is the essential factor to a team being successful compared with unsuccessful teams. Although no single factor leads to a team's success, the composite of the team's ability to be knowledgeable of, to be trained on, and to monitor and manage each of the core processes and emergent factors leads to their success.

Influencing Conditions

Beyond the core processes and emergent factors already identified, the following influencing conditions also are essential:

- context,
- composition, and
- culture.

The main difference between core processes and influencing conditions is that team members have the ability to manage and alter the core processes and emergent factors, whereas they have little to no control over the influencing conditions.

Context

As previously discussed, teams are contextual. This is evident in the variety of team types that are found within organizations and institutions. Teams vary in their structure, type, and interaction styles (e.g., project based, management, face to face, virtual). These different team types are formed for a specific function to address organizational problems. A team that is successful for one organizational setting may not be as successful in a different organizational setting. This is the general idea of context. For example, an emergency response team would not be the best team type for correcting a cavity at the dentist.

Composition

The composition relates to the previous discussion on team composition. Team member KSAOs for one organization may differ for another organization. Team composition depends on the context and setting expected of the team. Context relates to situational characteristics or external influences that affect a team. Context is important because "it can shape the very nature in which team members interact with one another" (Dihn and Salas, 2017: 28).

Culture

The last influencing condition is culture. Culture refers to the "assumptions people hold about relationships with each other and the environment that are shared among an identifiable group of people" (Dihn and Salas, 2017: 29). Although team composition also could concentrate on providing a diversified membership, being inclusive to all demographic characteristics (e.g., gender, race, ethnicity) of team members, once the team has formed, the team members have little to no control over changing the cultural characteristics of the team.

The team members must function with the team that has been put together. This relates to team members having little to no influence over culture.

Having little control over these influencing conditions highlights the importance that should be placed on leadership when considering team makeup, considering each teams' context, composition, and culture. Leadership within the context of a team is also highly dependent on culture.

Team Leadership and Culture

A team's leadership is a product of the team's culture. Research has shown that leadership styles are in congruence with the expectations of the local culture. Dorfman et al. (2012) found that leaders operating in societies that favored participatory leadership acted in participatory manners, societies that favored humane leadership traits acted in more humane ways, and societies requiring autonomy resulted in leaders operating using self-protective styles (Dorfman et al., 2012). As culture partially dictates the style of a leader, the same can be said of team leadership. A team's leadership is a product of the team's culture.

When viewing team culture, one must always keep the following items in mind:

- Remain cognizant of the practical implications related to cultural differences. For the unwary research team, power distance and uncertainty avoidance cultural dimensions can be cultural traps. High power distance cultures (or those socialized into a high power distance culture) will expect deference due to status differences. Cultures varying in uncertainty avoidance will find team differences regarding the implementation of deadlines, organizational structure issues, and stress level differences when the best made plans go astray.
- Continually remind team members (at least once a year) about *their* particular or peculiar cultural differences. We remain humble about recommending a mechanism to accomplish this. It would be best to have an interactive discussion that covers topics such as "here is how my culture differs from yours with respect to the task at hand," and maybe most importantly, "here is what we need to agree on." As just mentioned, the construct of "deadlines" would be a clear candidate for continuing discussion.
- Retain a good sense of humor. Realize at both emotional and intellectual levels that cultures really do differ. (Dorfman et al., 2012: 518)

Conclusion

Researcher's from Google's Aristotle program identified: "*Who* is on the team matters less than how the team members interact, structure their work, and view their contributions" (Rozovsky, 2015: first para.). The following essential characteristics, or dynamics, of successful teams, also identified as high-performance teams, for Google included the following:

1. Psychological safety: Can we take risks on this team without feeling insecure or embarrassed?
2. Dependability: Can we count on each other to do high quality work on time?
3. Structure & clarity: Are goals, roles, and execution plans on our team clear?
4. Meaning of work: Are we working on something that is personally important for each of us?
5. Impact of work: Do we fundamentally believe that the work we're doing matters? (Rozovsky, 2015: 4th para.)

These findings from Google's Aristotle program provides an excellent example for context, Google found that these five characteristics led to successful teams for them, for their team types, and for the work that they do. Looking at research from a broader perspective, the research presented in this chapter identified nine teamwork dimensions that have been found to lead to successful teamwork. These factors could be applied to many different organizations, contextual settings, and team types. These factors were classified as being core processes and emergent factors (cooperation, conflict, coordination, communication, coaching, cognition) or influencing conditions (context, composition, culture). The core processes and emergent factors are primarily internal dimensions of which team members must be cognizant and must be able to manage among themselves. The influencing conditions are primarily the responsibility of leadership, and leaders must consider the context, composition, and culture of each team.

In summary, effective teamwork consists of a whole lot more than just assembling a group of task experts (e.g., subject matter experts, SME); it also requires the knowledge of general teamwork skills (Dihn and Salas, 2017). Both taskwork and teamwork skills must be part of an organizations' training and development program, especially for those organizations that already are utilizing team-based structures and for those planning to transition to team-based structures. Training needs to take two paths. When training to develop taskwork skills, providing training for individual team members to attend is sufficient (Mathieu et al., 2008). Alternatively, when training to develop teamwork skills, or those behaviors designed toward effective team functioning, training is most effective when it is delivered to intact teams as opposed to training individual team members (Mathieu et al., 2008). Research has shown that teams that train together have better opportunities to practice their skills before applying them to real-world problems, allowing team members with "opportunities for members to integrate their teamwork skills and to jointly practice complex coordinated actions" (Mathieu et al., 2008: 447).

Only when teamwork skills begin to be developed for an organization's teams, in which team members are able to monitor and manage their own team's teamwork skills, can leadership then begin to focus on team effectiveness. Without effective teamwork and taskwork skills, a team's effectiveness will remain substandard at best. The next chapter will discuss the details of team effectiveness.

References

Amason AC. (1996) Distinguishing the effects of functional and dysfunctional conflict on strategic decision making: Resolving a paradox for top management teams. *Academy of Management Journal* 39: 123–148.

Behfar KJ, Mannix EA, Peterson RS, et al. (2010) Conflict in small groups: The meaning and consequences of process conflict. *Small Group Research* 42: 127–176.

Bolinger AR, Klotz AC, and Leavitt K. (2018) Contributing from inside the outer circle: The identity-based effects of noncore role incumbents on relational coordination and organizational climate. *Academy of Management Review* 43: 680–703.

Cohen SG and Bailey DE. (1997) What makes teams work: Group effectiveness research from the shop floor to the executive suite. *Journal of Management* 23: 239–290.

Cordery JL and Tian AW. (2017) Team design. In: Salas E, Rico R, and Passmore J (eds) *The psychology of team working and collaborative processes*. Malden, MA: Wiley, 105–128.

Denning S. (2018) *The age of Agile: How smart companies are transforming the way work gets done*. New York, NY: AMACOM.

Devine DJ. (2002) A review and integration of classification systems relevant to teams in organizations. *Group Dynamics: Theory, Research, and Practice* 6: 291–310.

Dihn JV and Salas E. (2017) Factors that influecne teamwork. In: Salas E, Rico R, and Passmore J (eds) *The Wiley Blackwell handbook of the psychology of team working and collaborative processes*. Malden, MA: Wiley, 15–41.

Dorfman P, Javidan M, Hanges P, et al. (2012) GLOBE: A twenty year journey into the intriguing world of culture and leadership. *Journal of World Business* 47: 504–518.

Duhigg C. (2016, February 25) What Google learned from its quest to build the perfrect team. *New York Times Magazine*.

Edmondson A. (1999) Psychological safety and learning behavior in work teams. *Administrative Science Quarterly* 44: 350–383.

Edmondson AC. (2019) *The fearless organization: Creating psychological safety in the workplace for learning, innovation, and growth*. Hoboken, NJ: Wiley.

Elverum C and Welo T. (2016) Leveraging prototypes to generate value in the concept-to-production process: a qualitative study of the automotive industry. *International Journal of Production Research* 54: 3006–3018.

Falk-Krzensinski HJ. (2012) Guidance for team science leaders: Tools you can use. *The Academic Executive Brief*. Available at: http://academicexecutives.elsevier.com/articles/guidance-team-science-leaders-tools-you-can-use.

Falk-Krzesinski HJ, Contractor N, Fiore SM, et al. (2011) Mapping a research agenda for the science of team science. *Research Evaluation* 20: 145–158.

Fiore SM. (2008) Interdisciplinarity as teamwork: How the science of teams can inform team science. *Small Group Research* 39: 251–277.

Forsgren N, Humble J, and Kim G. (2018) *Accelerate: Building and scaling high performing technology organizations*. Portland, OR: IT Revolution.

Forsyth DR. (2014) *Group dynamics*. Belmont, CA: Cengage Learning.

Greer LL and Dannals JE. (2017) Conflict in Teams. In: Slalas E, Rico R, and Passmore J (eds) *The Wiley Blackwell handbook of the psychology of team working and collaborative processes*. Malden, MA: Wiley Online Library, 317–343.

Hackman RJ. (2011) *Collaborative intelligence: Using teams to solve hard problems*. San Francisco, CA: Berrett-Koehler.

Hollenbeck JR, Moon H, Ellis APJ, et al. (2002) Structural contingency theory and individiual differences: Examination of external and internal person-team fit. *Journal of Applied Psychology* 87: 599–606.

Jehn KA and Chatman JA. (2000) The influence of proportional and percetpuatl conflict composition on team performance. *International Journal of Conflict Management* 11: 56–73.

Jehn KA and Shah PP. (1997) Internpersonal relationships and task performance: An examination of mediating processes in friendship and acquaintance groups. *Interpersonal Relations and Group Processes* 72: 775–790.

Karlgaard R and Malone MS. (2015) *Team Genius: The new science of high-performing organizations.* New York, NY: Harper Collins.

Katzenbach JR and Smith DK. (2001) *The discipline of teams: A mindbook-workbook for delivering small group performance.* New York, NY: Wiley.

Kayser T. (2011) *Building team power: How to unleash the collaborative genius of teams for increased engagement, productivity, and results.* New York, NY: McGraw-Hill Professional.

Kozlowski SWJ, Grand JA, Baard SK, et al. (2015) Teams, teamwork, and team effectiveness: Implications for human system integration. In: Boehm-Davis D, Durso F and Lee J (eds) *APA handbook of human systems integration.* Washington, DC: APA, 555–571.

Kozlowski SWJ and Ilgen DR. (2006) Enhancing the effectiveness of work groups and teams. *Psychological Science of Work Groups and Teams* 7: 77–124.

Lewis K. (2003) Measuring transactive memory systems in the field: Scale development and validation. *Journal of Applied Psychology* 88: 587–604.

Lim B-C and Klein KJ. (2006) Team mental models and team performance: A field study of the effects of team mental model similarity and accuracy. *Journal of Organizational Behavior* 27: 403–418.

Lord RG, Gatti P, and Chui SLM. (2016) Social-cognitive, relational, and identity-based approaches to leadership. *Organizational Behavior and Human Decision Processes* 136: 119–134.

Marks MA, Mathieu JE, and Zaccaro SJ. (2001) A temporally based framework and taxonomy of team processes. *Academy of Management Review* 26: 356–376.

Mathieu JE, Heffner TS, Goodwin GF, et al. (2005) Scaling the quality of teammates' mental models: Equifinality and normative comparisons. *Journal of Organizational Behaivor* 26: 37–56.

Mathieu JE, Maynard TM, Rapp T, et al. (2008) Team effectiveness 1997-2007: A review of recent advancements and a glimpse into the future. *Journal of Management* 34: 410–476.

Mayer RC, Davis JH, and Shoorman DF. (1995) An integrative model of organizational trust. *Academy of Management Review* 20: 709–734.

McIntyre RM and Salas E. (1995) Measuring and managing for team performance: Emerging principles from complex environments. In: Guzzo RA and Salas E (eds) *Team effectiveness and decision making in organizations.* San Francisco, CA: Jossey-Bass.

Mohammad A, Varun G, and Thatcher JB. (2015) Taxing the development structure of open source communities: An information processing view. *Decision Support Systems* 80: 27–41.

National Cancer Institute. (n.d.) *About team science.* Team Science Toolkit. Available at: https://www.teamsciencetoolkit.cancer.gov/Public/WhatIsTS.aspx.

Rico R, Sanchez-Manzanares M, Gil F, et al. (2008) Team implicit coordination processes: A team knowledge-based approach. *Academy of Management Review* 33: 163–184.

Rigby DK, Sutherland J, and Noble A. (2018) Agile at scale: How to go from a few teams to hundreds. *Harvard Business Review* 96: 88–96.

Rozovsky J. (2015, November 17) *The five keys to a successful Google team.* Available at: https://rework.withgoogle.com/blog/five-keys-to-a-successful-google-team/.

Salas E, Rico R, and Passmore J. (2017) The psychology of teamwork and collaborative processes. In: Salas E, Rico R, and Passmore J (eds) *The Wiley Blackwell handbook of the psychology of team working and collaborative processes.* Malden, MA: Wiley.

Salas E, Sims DE, and Burke SC. (2005) Is there a "big five" in teamwork? *Small Group Research* 36: 555–599.

Slyngstad DJ, DeMichele G, and Salazar MR. (2017) Team performance in knowledge work. In: Salas E, Rico R, and Passmore J (eds) *The Wiley Blackwell handbook of the psychology of team working and collaborative processes.* Malden, MA: Wiley, 43–71.

Stokols D, Hall KL, Taylor BK, et al. (2008) The science of team science: Overview of the field and introduction to the supplement. *American Journal of Preventive Medicine* 35: S77–S89.

Thompson LL. (2018) *Making the team: A guide for managers.* New York, NY: Pearson.

Turner JR. (2016) Team cognition conflict: A conceptual review identifying cognition conflict as a new team conflict construct. *Performance Improvement Quarertly* 29: 145–167.

Weaver SJ, Feitosa J, Salas E, et al. (2013) The theoretical drivers and models of team performance and effectiveness for patient safety. In: Salas E and Frush K (eds) *Improving patient safety through teamwork and team training.* New York, NY: Oxford University Press, 3–26.

Wildman JL, Thayer AL, Rosen MA, et al. (2012) Task types and team-level attribues: Synthesis of team classification literature. *Human Resource Development Review* 11: 97–129.

Wolfson MA and Mathieu JE. (2017) Team composition. In: Salas E, Rico R, and Passmore J (eds) *The Wiley Blackwell handbook of the psychology of team working and collaborative processes* Maldin, MA: Wiley Blackwell, 129–149.

Wu L, Wang D, and Evans JA. (2019) Large teams develop and small teams disrup science and technology. *Nature* 566: 378–382.

Yasmi Y, Schanz H, and Salim A. (2006) Manifestation of conflict escalation in natural resource management. *Environmental Science and Policy* 9: 538–546.

CHAPTER 11

Team Effectiveness

Teams that have been shown to be effective include those that are knowledge-able and capable of combining both processes of teamwork and taskwork. Teams are not easily formed (composed) or put into practice. Teams composed solely of taskwork skills, that is, of individuals capable only of completing the tasks required of a team, with no teamwork skills or competencies, have been shown to be ineffective (Christodoulou et al., 2008). This point has been high-lighted by others: "Based on antidotal experiences with teams and empirical or theoretical support, it is known that teams are not easily implemented, that the creation of a team of skilled members does not ensure success, and that team-work does not just happen" (Salas et al., 2005: 556).

Team effectiveness has been mistaken in the literature as being team perfor-mance, it also has been represented using many different meanings with no univer-sally accepted definition (Guzzo and Dickson, 1996). Team effectiveness has been referred to as the point at which team processes are aligned with task demands (Kozlowski et al., 2015) and is considered optimized when the processes produce the desired outcome (providing value to the customer) (Driskell et al., 2018).

The effectiveness of teams originally was identified as having three compo-nents: "group performance, satisfaction of group-member needs, and the ability of the group to exist over time" (Gladstein, 1984: 500; see also Hackman and Morris, 1975). These three dimensions have been used to evaluate the effective-ness of teams in practice and have been improved on over the years. These three revised dimensions, provided by Hackman in studying intelligent teams that often are operating in the complex domain, include the following:

1. *The productive output of the team (that is, its product, service, or decision) meets or exceeds the standards of quantity, quality, and timeliness of the team's clients—that is, of the people who receive, review, or use the output.*

2. *The social processes the team uses in carrying out the work enhance members' capability to work together interdependently in the future.*
3. *The group experience, on balance, contributes positively to the learning and professional development of individual team members.* (Hackman, 2011: 37–39; emphasis in original)

Team effectiveness, according to Hackman, results in customers' view of the team's product or outcome. This reemphasizes the philosophy of customer first and the concept of *flow* by providing value to the customer. A team's output is tied directly to the customer. The social processes involve teams that are capable of performing successful teamwork activities to produce the final outcome expected from the customer. The last component, team member experience, involves the affective component of being a member of a team and an organization (e.g., past experiences, current experiences, team learning). Did the team members gain from the experience, did they find purpose in their work while being connected to the team, and did the team experience aid in each team member's developmental efforts? Earlier we discussed the role of leadership in instilling purpose in the work of their followers. Without providing this purpose, leadership will fail to provide a meaningful experience for team members. With purpose, the team and team members excel: "The benefits of teamwork come only when capable people work together interdependently to achieve some collective purpose" (Hackman, 2011: 29).

Team effectiveness is different from team performance. Team effectiveness relates to a team's outcome (performance) as well as to the interactions (teamwork) and processes (teamwork and taskwork) used to produce an outcome. Team performance primarily focuses on a team's outcome, regardless of the teamwork or taskwork processes to get to that outcome (Salas et al., 2005). Team effectiveness is a holistic perspective viewing a team's performance as well as the teamwork and taskwork processes used to achieve the team's outcome (Salas et al., 2005).

Monitoring a team's progress can help determine how successful that team will be. The following three general team aspects are predictive of a team's success:

1. The amount of *effort* members are expending in carrying out their collective work.
2. The task-appropriateness of the team's *performance strategies*; the choices the team makes about how it will carry out the work.
3. The level of *knowledge and skill* the team is applying to the work. (Hackman, 2011: 40)

Team Effectiveness Frameworks

The literature has provided a number of team effectiveness frameworks. The most common is McGrath's input-process-output (IPO) framework (McGrath, 1964), followed by the input-mediator-output-input (IMOI) framework (Ilgen et al., 2005; Mathieu et al., 2008), the input-process-emergence-output (IPEO) framework (Turner and Baker, 2017), a team process model the incorporates processes and emergent factors as a mediator (Magpili and Pazos, 2018), and team transition processes (Marks et al., 2001; see Chapter 10). Each of these frameworks are briefly described next.

The IPO Framework

A common framework used to identify the components for team effectiveness is the IPO framework shown in Figure 11.1, originally presented by McGrath (1964). Using systems theory, inputs represent antecedents that contribute to the process, and the process produces the final outcome. When viewing teams from the IPO perspective, antecedents either enable or constrain team member interactions (Mathieu et al., 2008; Mathieu et al., 2019). Processes transform the antecedents, team member interactions, for example, to outcomes, also known as the "by-products of team activity" (Mathieu et al., 2008: 412).

The literature has defined the following inputs, processes, and outcomes relating to the IPO framework.

Inputs

Inputs at the team member level are related to individual characteristics: competencies, personalities, KSAs, and demographics (Kozlowski et al., 2015).

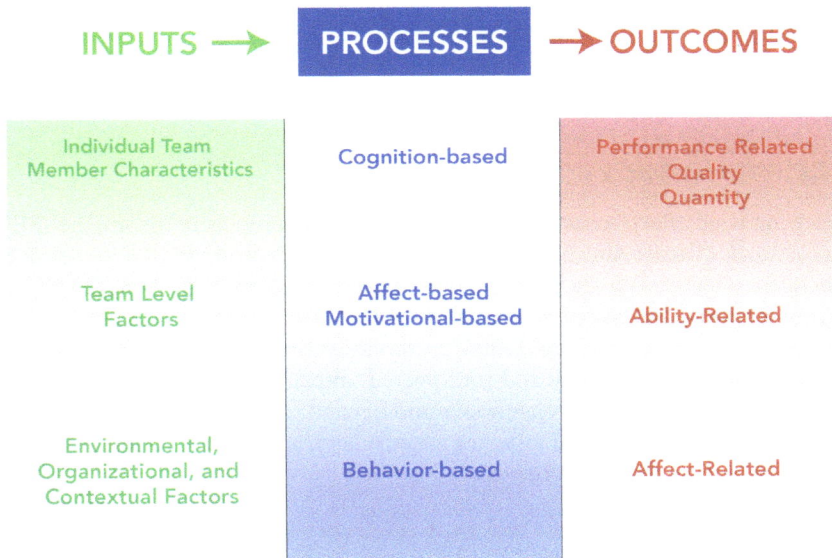

FIGURE 11.1.   The IPO Framework
Source: Adapted from McGrath (1964).

Team-level factors include examples, such as task structure, external leader influences, size, and power structure (Kozlowski et al., 2015). Organizational and contextual factors include organizational design features and environmental complexity, and environmental factors include external stressors and reward conditions (Kozlowski et al., 2015).

Processes

Processes include cognitive, affective, and behavioral phenomena that emerge from team member interactions; these processes also influence team outcomes. Examples of processes include developing transactive memory systems, cohesion-building, and collaborative problem solving (Kozlowski et al., 2015).

Outcomes

Outcomes result from team activity, reflecting the collective efforts of the team. Examples of outcomes can be performance based (i.e., quality, quantity), or they can be affective based (i.e., team member satisfaction, commitment) (Kozlowski et al., 2015; Mathieu et al., 2019).

The IMOI Framework

An extension to the IPO framework includes the IMOI framework (Ilgen et al., 2005; Mathieu et al., 2008). Rather than looking at processes, team-mediating mechanisms are the focal point in IMOI models (Mathieu et al., 2019). A primary advantage of this model is that mediators (the M in the IMOI model) provide a broad coverage of emerging states compared with processes. The I in the IMOI model represents a continuous process by returning back to the input.

The IPEO Framework

Viewing team effectiveness from the perspective of complexity theory, the IPEO framework (Turner and Baker, 2017) introduced the property of emergence to the team effectiveness framework literature. Emergence results from the "interaction, tension, and exchange rules governing changes in perceptions and understandings" (Lichtenstein et al., 2006: 2). Inputs are introduced to team processes that later emerge into new and unexpected outcomes, setting teams that pass through the stage of emergence above and beyond teams that do not experience emergence. Facilitating and fostering a team's environment and its interactions is essential to developing emergence.

The IPOI Framework

One iteration of the IPO framework combines processes and emergent states (Magpili and Pazos, 2018). Inputs are distinguished among individuals, the team, and organizational antecedents that lead into the process–emergent phase. The process results in team outcomes with feedback mechanisms that create an input-process–emergent-outcome-input (IPOI) framework.

Team Transition Processes

Team transition processes (Marks et al., 2001) describe the temporal processes that teams experience from a team's composition to the end of its tenure. The taxonomy of team processes includes transition phase processes in which the team focuses on team goals, the action phase processes in which team members focus on goal accomplishment, and the interpersonal processes in which team members manage interpersonal relationships (Marks et al., 2001).

TABLE 11.1.   Team Coevolving Model

Structural Features (Inputs)	Compositional Features (Inputs)	Mediating Mechanisms (Mediators)
Interdependence	Demographic, functional, and personality diversities	Action, transition and interpersonal processes
Structural contingencies	Faultlines	Cohesion
Task scope, complexity, and structure	Member ability	Conflict
Technology	Member churn	Creativity
Virtuality		Information sharing
		Motivation
		Trust
	Transactive memory systems	Transactive memory systems
	Shared cognition	Shared cognition
	Psychological safety	Psychological safety
Differentiation	Differentiation	
Member centrality	Member centrality	
Roles	Roles	
Skill and authority	Skill and authority	
Adaptability		Adaptability
Boundary spanning		Boundary spanning
Empowerment		Empowerment
Shared leadership		Shared leadership
Team effectiveness	Team effectiveness	Team effectiveness
Individual reactions	Individual reactions	Individual reactions
Learning	Learning	Learning

Source: From Mathieu et al. (2019).

Temporal Dynamics and Team Effectiveness

[handwritten margin note: Too theoretical for me]

Team effectiveness frameworks (such as IPO) combined with the temporal dynamics presented in the team transition process (Marks et al., 2001) extend the current research as described in the following:

> We adopt a more contemporary perspective that has evolved over the last decade, which conceptualizes the team as embedded in a multilevel system that has individual, team, and organizational-level aspects; which focuses centrally on *task-relevant processes*; which incorporates *temporal dynamics* encompassing episodic tasks and developmental progression; and which views team processes and effectiveness as *emergent phenomena* unfolding in a proximal task or social context that teams in part enact while also being embedded in a larger organization system or environmental context. (Kozlowski and Ilgen, 2006: 80)

[handwritten margin note: The paper is too complex for me]

The ability to incorporate temporal features with team effectiveness frameworks, while viewing teams as complex adaptive systems, has been presented as a modification of the IMO framework by Mathieu et al. (2019). Their team coevolving model provides "team inputs, mediating mechanisms, and structural features ... as overlapping coevolving facets of teams that collectively combine to generate effectiveness" (Mathieu et al., 2019: 19). This coevolving model was categorized into three components—that is, compositional features (Inputs), structural features (Inputs), and mediating mechanisms (Mediators)—with three external influences (i.e., organizational structure and culture, multiteam features, and external leadership).

Our research incorporates the team transition phases (transition, action, interpersonal processes (Marks et al., 2001) by separating processes and mediating factors among individual, team, and multiteam system (MTS[1]) levels of analysis in an MTS effectiveness model (Turner et al., 2020). This MTS level could be replaced with an organization if only one network of teams is involved. When multiple networks of teams are involved, there is an additional level—that is, individual, team, MTS, and organization. The levels of analysis are contextual and depend on the organization's structure. Table 11.2 provides the processes and mediating factors that lead to effective teams.

The processes and mediating factors identified in Table 11.2 can be broken down into three levels: individual, team, and MTS. The processes and mediating factors at the individual level can be used not only to monitor and evaluate existing teams but also to identify specific individual competencies for development purposes. This MTS effectiveness model provided in Table 11.2 also includes different characteristics for each of the team temporal phases (transaction, action, interpersonal). Development must focus on the individual first, then the

---

1 Multiteam systems (MTS) will be covered in more detail in Chapter 12. Basically, MTS is when teams are scaled, also known as team of teams, or scrum of scrums.

TABLE 11.2. MTS Effectiveness Model

Temporal Processes with Levels of Analysis		
MTS	*Team*	*Individual*
Transition Phase Processes: Mission Analysis Formulation and Planning		
Closed-loop communication	Mission analysis	
Duration	Planning	
Genesis	Team mission	
Hierarchical arrangement		
Leadership—external		
Leadership—boundary spanning		
Network		
Organizational structure		
Planning		
Shared leadership		
Transition Phase Processes: Goal Specification		
National culture	Accurate problem models	
Organizational culture	Task characteristics	
Organizational diversity	Team goal commitment	
Organizational goals		
Planning		
Transition Phase Processes: Strategy Formation		
Collaboration with interagency partners	Developing strategy	
Communication structure	Finishing	
Functional diversity		
Geographically dispersion of teams		
Innovation		
Leader strategizing		
Mode of communication		
Organizational diversity		
Organizational policies		
Power distribution		
Stage of development		
Temporal orientation		
Training		
Transformation of development		
Understanding of MTS couplings		
Action Phase Processes: Monitoring Progress Toward Goals		
Decision making	Cue-strategy associations	Leadership (shared)
	Mutual performance monitoring	Task-related assertiveness
	Problem detection	
	Workload sharing	

(Continued)

TABLE 11.2. MTS Effectiveness Model (Continued)

Temporal Processes with Levels of Analysis		
MTS	**Team**	**Individual**
Action Phase Processes: Systems Monitoring		
Intrateam feedback	Gathering information	
Resources		
Action Phase Processes: Teams Monitoring and Backup Responses		
Brainstorming	Adapting	Openness to change
Reprioritizing of goals	Backup supporting behavior	
	Helping	
	Learning from team's best member	
	Team orientation	
	Workload sharing	
Action Phase Processes: Coordination Activities		
Coordination—horizontal and vertical	Coordination horizontal	Individual roles
Coordination for readiness	Implicit coordination strategies	Planning
Leader coordination	Interdependence	Self-management skills
	Planning	Work experience
	Team autonomy	
Interpersonal Phase Processes: Conflict Management		
Conflict management		
Risk mitigation		
Structuring		
Interpersonal Phase Processes: Motivation and Confidence Building		
Motivation of others		
Motive structure		
Team efficacy		
Team learning		
Team rewards		
Interpersonal Phase Processes: Affect Management		
Climate for information sharing	Mutual trust	Individual autonomy
Cohesion	Peer control	General skills
Information sharing	Shared mental models	Trust
Managing diversity	Social identity	
Psychological safety	Team cohesion	
Social identity	Team empowerment	
Team-level diversity	Team learning orientation	
Team psychological safety	Team potency	
	Team psychological safety	
	Team satisfaction	

Source: Transition phase processes from Marks et al. (2001). MTS team effectiveness model components from Turner et al. (2020).

team level, and last the MTS and organizational levels. The processes and mediating factors provided in Table 11.2 identify essential factors that should be practiced to ensure successful team effectiveness outcomes. This MTS effectiveness model provides leaders with the tools needed to implement adaptable characteristics into their teams, especially when involving MTS. The next chapter will cover MTS in more detail.

Team Effectiveness Formulae

To summarize the material in this chapter thus far, team effectiveness involves a team's performance as well as the teamwork and taskwork processes that provide value to the customer. Team effectiveness (TE) is a factor of the following components: teamwork (TW), taskwork (TK), performance or outcome (PF), and value provided to the customer (CV). The formulae for team effectiveness is written as follows:

$$TE = TW + TK + PF + CV \text{ (see Turner et al., 2020)}$$

Chapter 10, Teams, Teamwork, and Taskwork, introduced the team transition phases. These transition phases included three processes: transition phase, action phase, and interpersonal processes (Marks et al., 2001). These three processes represent the different phases that effective teams go through during their tenure and as they work to achieve a team's objective or goal. Transition phases represent evaluating and planning activities toward achievement of a team's goal, action phases include activities toward achievement of a team's goal, and interpersonal phases identify the interpersonal (relationship) processes necessary to achieve a team's goal (Marks et al., 2001). From this perspective, the formulae also would need to include these three phases. In this revised formulae we identified three processes in the following manner: transition phases as TP, action phases as AP, and interpersonal phases as IP. This revised comprehensive formulae follows:

$$TE = (TW + IP) + TK(TP + AP) + PF + CV \text{ (see Turner et al., 2020)}$$

Team effectiveness is composed of teamwork as a function of both teamwork and interpersonal phases; taskwork is a function of transition and action phases; performance is a function of quality, quantity, and time characteristics; and customer value is a function of customer satisfaction as well as team member satisfaction and commitment resulting from team-member experiences. By combining the team transition phases (Table 10.1) and the teamwork core processes and emergent factors (Table 10.3), we can identify the constructs and variables for each of the formulae components (highlighted in Table 11.3).

TABLE 11.3.   Team Effectiveness Components

(TW + IP)	TK (TP + AP)	PF	CV
Teamwork	Taskwork (Transition Phase Processes + Action Phase Processes)	Performance	Customer value (internal, external, or both)
Team composition	Orientation	Quality	Feedback
Cooperation	Resource coordination	Quantity	Satisfaction
Goal commitment	Timing	Time	Exceeds expectations
Checking in	Response coordination		Satisfaction of team members
Psychological safety	Motivation		Team member commitment
Trust	Systems monitoring		
Cohesion	Procedures maintenance		
Conflict	**Transition Phase Processes**		
Cognition conflict	–Mission analysis		
Process conflict	–Goal specification		
Status conflict	–Strategy formulation and planning		
Task conflict	–Deliberate planning		
Coordination	–Contingent planning		
Explicit	–Reactive strategy adjustment		
Implicit	**Action Phase Processes**		
Communication	–Monitoring progress toward goals		
Coaching	–Systems monitoring		
Cognition	–Internal systems monitoring		
Shared cognition	–Environmental monitoring		
Transactive memory system	–Team monitoring and backup responses		
Shared mental memory	–Coordination activities		
Task-related features			
Team-related aspects			
Culture			
Context			
Leadership structure			
Interpersonal Phase Processes			
Conflict management			
Preemptive conflict management			

TABLE 11.3.   Team Effectiveness Components (Continued)

(TW + IP)	TK (TP + AP)	PF	CV
Teamwork	Taskwork (Transition Phase Processes + Action Phase Processes)	Performance	Customer value (internal, external, or both)
Reactive conflict management			
Motivation and confidence building			
Affect management			

Sources: Adapted from Behfar et al., (2010); Devine (2002); Dihn and Salas (2017); Edmondson (1999); Edmondson (2019); Forsyth (2014); Greer and Dannals (2017); Hackman (2011); Jehn and Chatman (2000); Lewis (2003); Lim and Klein (2006); Marks et al. (2001); Mathieu et al. (2019); Mathieu et al. (2005); Mathieu et al. (2008); Mayer et al. (1995); Rico et al. (2008); Turner (2016); Weaver et al. (2013); Yasmi et al. (2006).

Team Effectiveness Training

To sustain productive teamwork behaviors, team training must provide team members with monitoring, regulating, and self-corrective actions to keep them on the correct path (Salas et al., 2015). Teams must undergo a series of continuous training and development initiatives and programs, from initial team-based training to in situ coaching and self-regulating behaviors, to debriefing and reflection exercises after critical stages of each team's task episode. Team training features two concepts: (1) "team competencies" and (2) "instructional strategies for teaching these competencies" (Alonso et al., 2006: 401). Team competencies are contextual and must be relevant to the workplace setting and task. Instructional strategies need to evaluate learning from the training to show that deep learning has been achieved, which often is reflected in the team's daily work activities.

Many problems with today's workplace training can be summarized into three problem areas. First, training is often conducted at the individual level. Individual team members may attend training at various times, separately, until each team member eventually has attended and received training. In some cases, only a few team members receive training. Those who do receive the training are then expected to transfer what was learned to other team members. Although this method of training may be the more convenient means of training, it misses the essential aspect of teamwork. Team members are expected to learn to function together, to learn together, and to provide oversight mechanisms to correct or eliminate errors, but individual training does not include any of these aspects of teamwork.

Training, individually, is just that, individual training. It does not account for, or take advantage of, building and enhancing existing teamwork capabilities of team members. Research has shown that teams that train together perform their tasks better than teams whose members train individually (Liang et al., 1995;

Salas et al., 2015). Research has highlighted the fact that teams that trained together were more able to "(a) recall different aspects of the task, (b) coordinate their task activities, and (c) trust one another's expertise" (Liang et al., 1995: 390). Training together supports team members in building shared knowledge, enhancing team cognitive functions, and developing coordination activities among members. Team training "builds shared mental models of the situation, task environment, and interactions of team members increases a team's ability to function effectively under high levels of stress" (Salas et al., 2008: 542).

The second common problem with training, in general, is that it is often a futile exercise. Training is often the course of action taken when problems persist and are misunderstood. It is easier to assign training than to identify the problem, develop a course of action to address the possible causal mechanisms to the problem, and evaluate the effectiveness of the performance improvement intervention used to solve the problem. Training is required for one primary purpose: to address a deficit in knowledge or skills. When there is a human performance (P) problem, one can look to two potential mechanisms: behavior (B) or the environment (E; $P = B \times E$) (Gilbert, 2007). Performance can be broken down into an individual's behavior, including knowledge, skills, abilities, relationships, and interactions. It also can be broken down into the environment, including incentives, tools, and resources, and the organizational structure. If the problem resides in the behavioral category, then training may be a valid solution. If, however, the problem resides in the environmental category, training is insufficient.

The third issue with training is that it mainly concentrates on developing skills to achieve taskwork while often ignoring teamwork behaviors: "The majority of this training is based around the development of technical skills specific to crew-member roles and tasks, with limited time devoted to the development or more generalizable teamwork skills" (Salas et al., 2015: 204). Taskwork training is essential when there is an identifiable deficit in knowledge surrounding a specific skill. When dealing with team-based structures, however, it is essential to develop teamwork skills surrounding taskwork in conjunction with skill-based training. Training the team as a whole will develop teamwork skills while also providing cross-training for team members, gaining new and improved knowledge in areas outside of their current domain of expertise.

Team training works. When team members train as a cohesive unit, it promotes "teamwork and enhances team performance" (Salas et al., 2008: 542). Research has captured the benefits of team training, not only training as a team but also training teamwork capabilities during real-time activities. Before beginning any team training initiatives, it is essential to first clearly identify the essential teamwork skills necessary for the contextual setting. Teamwork is contextual and varies across settings. Training programs should be designed around those essential teamwork skills. The teamwork skills presented in Table 11.3 provide a starting point, but these teamwork skills need to be identified for each contextual setting. A few productive examples of team training for teamwork skills, from the literature, are provided in the following sections.

TABLE 11.4.   Team Dimensional Training Teamwork Behaviors

Teamwork Dimension	Teamwork Behaviors
Information Exchange	Sharing information to the right teammate at the right time; seeking information from all relevant sources; providing periodic situational update; summarizing the big picture.
Communication Delivery	Utilizing a shared lexicon and terminology; avoiding excess chatter; speaking clearly; reporting in appropriate order.
Support Behavior	Offering, requesting, accepting, correcting.
Initiative and Leadership	Explicitly stating priorities; providing guidance; making suggestions; directing team members.

Source: Smith-Jentsch et al. (2008).

Team Dimensions Training

Self-correction activities among team members has been shown to improve teamwork capabilities, taskwork skills, and coordination. One technique that utilizes self-correction activities among team members is the team dimensional training (TDT) program. Team dimensional training was conducted for a Naval exercise in which 11 teamwork behaviors were identified and categorized according to four higher-order dimensions: information exchange, communication delivery, supporting behavior, and initiative and leadership (Smith-Jentsch et al., 2008). These dimensions and teamwork behaviors are presented in Table 11.4 and include those teamwork behaviors that were relevant to the contextual setting (i.e., the Naval task and exercise).

TDT measured shared mental models to identify two components: the similarity among team members and the accuracy of the information compared with some standard or benchmark level. The model presented in Table 11.4 was used not only to describe how effective teams behaved but also to provide a framework for team members to diagnose and solve their performance problems (Smith-Jentsch et al., 2008).

Self-regulation involves briefing (a priori) and debriefing[2] (post hoc) activities. Research has shown that debriefing activities can improve team effectiveness by approximately 25% (Tannenbaum and Cerasoll, 2012). Briefing and debriefing activities are successful when a framework is utilized to avoid the potential of team members developing inaccurate shared mental models (Smith-Jentsch et al., 2008). The framework developed specifically for this Naval exercise was used as a tool for the briefing and debriefing exercises, which provided a context-specific framework to guide these processes.

---

2   Debriefing involves team members reflecting on their experiences so the team can come to an agreement on what changes are necessary to perform the team's task better (Salas et al., 2015).

The self-correction aspect of the TDT involved the model that was developed for the Naval exercise for use during briefing and debriefing sessions. This process involved a facilitator asking the team "to describe both positive and negative instances of their own performance that illustrate each component within the expert model [Table 11.4] and, on the basis of this discussion, identify and agree upon process-oriented goals for improvement" (Smith-Jentsch et al., 2008: 312).

Results indicated that team effectiveness improved for teams that utilized the expert model during the briefing and debriefing sessions compared with those that did not use the expert model. Results were apparent after two training cycles, which showed that shared mental models were being developed and contributed positively to the team's effectiveness.

Developing an empirical model of teamwork behaviors would be required before using TDT. Given that teamwork is contextual depending on the team type, type of task, and organizational setting, an empirical model of teamwork behaviors is essential before beginning any training program to ensure that the correct teamwork behaviors are being developed. Briefing and debriefing exercises also are essential because they represent the reflection process of learning, thus allowing for modifications to improve overall team effectiveness.

Simulation-Based training

Simulation-based training (SBT) is one method that has been shown to be a powerful training mechanism to improve team effectiveness. SBT has the advantage of allowing teams "to engage in the dynamic social, cognitive, and behavioral processes of teamwork and receive feedback and remediation based on team performance" (Salas et al., 2008: 542). Training a team as a whole, and training them in a dynamic setting, forces team members to interact. This interaction reflects the team's teamworking capabilities. By observing teams during simulations, a team's failures or shortcomings in teamwork can be identified. Once identified, teamwork skills can be trained through coaching, briefing and debriefing exercises, and repetitive exercises that force team members to interact and practice new skills to reinforce learning.

Although numerous simulation-type training modules are available from which to choose, it is essential to ensure that the simulation used for team training is designed to cause the desired team member interactions. While team members go through the exercises in various modules, the appropriate teamwork skills need to be practiced and mastered before completing each module. The teamwork skills selected for the simulation modules should be designed around the specific teamwork skills that are necessary for the contextual setting. As in TDT, the essential teamwork behaviors for the team need to be identified and captured in each training module of the simulator. Research has provided evidence in support of SBT: "well-designed team training increases the quality of team processes and over-all performance outcomes" (Salas et al., 2008: 542).

Debriefing Activities   ⎰⸒ ⸗

Debriefing activities must be part of any training exercise, especially any SBT (Tannenbaum and Cerasoll, 2012). Debriefings can be used as a means of live training if conducted in the workplace on active teams. Debriefing exercises must involve all team members. Debriefings must focus on improving teamwork and taskwork skills, which requires the essential teamwork and taskwork skills to be identified in advance. These exercises must measure and evaluate the effectiveness of the team (Tannenbaum and Cerasoll, 2012). When incorporating debriefings, either in real team activities or in training activities, the following key points need to be considered:

- Debriefs are a quick, effective tool for improving team and individual performance.
- Meta-analytic results from 46 independent samples show that debriefs improve performance an average of 20% to 25%. Debriefs work equally well for teams and individuals.
- Debriefs work best when properly aligned: If the goal is to improve team performance, debriefs should be conducted with, measure, and focus on teams rather than individuals (and vice versa).
- Findings suggest that debriefs are even more effective when structured and facilitated. (Tannenbaum and Cerasoll, 2012: 242)

TeamSTEPP Training

The Team Strategies and Tools to Enhance Performance and Patient Safety (TeamSTEPPS) training, which focuses on core teamwork competencies and behaviors, has been developed for and utilized in for the healthcare sector (Alonso et al., 2006). The TeamSTEPPS training program has provided team members with the competencies to "monitor the performance of others and provide assistance, to plan and organize team roles, and to communicate with one another efficiently; combining these skills yield a highly adaptable and flexibly team" (Alonso et al., 2006: 406). These competencies also develop self-correcting behaviors in which team members support and correct each other in real-time.

Four teamwork skills are provided in the TeamSTEPPS training: leadership, situation monitoring, mutual support, and communication (Alonso et al., 2006). Table 11.5 presents each of these four skills, describes each skill, and provides examples of each behavior. Teamwork outcomes expected from the TeamSTEPPS training include mutual trust, adaptability, shared mental models, and team orientation (Alonso and D'unleavy, 2013).

The TeamSTEPPS training has been delivered in a number of different contextual settings through the Department of Defense (DoD) healthcare operations and the Health Care Team Coordination Program (HCTCP) to Army and Navy Military Health System (MHS) (Alonso et al., 2006). Team training efforts using

TABLE 11.5.   TeamSTEPP KSAs

Teamwork Skills	Description	Behaviors
Leadership	Ability to direct and coordinate the activities of other team members, assess team performance, assign tasks, develop team knowledge, skills, and attitudes, motivate team members, plan and organize, and establish a positive atmosphere.	• Facilitate team problem solving. • Provide performance expectations and acceptable interaction patterns. • Synchronize and combine individual team member contributions. • Seek and evaluate information that affects team functioning. • Clarify team member roles. • Engage in preparatory meetings and feedback sessions with the team.
Situation Monitoring	Ability to develop common understandings of the team environment and apply appropriate task strategies to accurately monitor teammate performance.	• Identify mistakes and lapses in other team members' actions. • Provide feedback regarding team members' actions to facilitate self-correction.
Mutual Support	Ability to anticipate other team members' needs through accurate knowledge about their responsibilities and to shift workload among members to achieve balance during high periods of workload or pressure.	• Achieve recognition by potential backup providers that there is a workload distribution problem in their team. • Shift work responsibilities to underused team members. • Complete the whole task or parts of tasks by other team members.
Communication	Ability to exchange information between a sender and a receiver irrespective of the medium.	• Follow up with team members to ensure the message was received. • Acknowledge that a message was received. • Clarify with the sender of the message that it was received is the same as the intention.

Source: Alonso et al. (2006); Alonso and D'unleavy (2013).

the TeamSTEPPS program can be summarized in the following way: "Teams who engage in effective leadership, situation monitoring, mutual support, and communication have better team outcomes such as shared mental models, mutual trust, team orientation, and adaptability" (Alonso and D'unleavy, 2013: 56).

Team Effectiveness Training Summary

Team effectiveness training must begin with acknowledgment of the requisite teamwork skills and competencies that are essential to the contextual setting. Success from the TeamSTEPPS training that was achieved in the healthcare field partially was attributed to efforts to first identify the team's knowledge, skills, and attitudes that were relevant to their environment. This was evident in the following: "Without a focus on these competencies, healthcare professionals can only attempt to realize safer care in an incomplete fashion" (Alonso and D'unleavy, 2013: 56). For team effectiveness training, teamwork skills must be identified first, these skills must be relevant to the contextual settings, and any training program must be focused on learning (content and activities) these teamwork activities. Training programs should be guided by the following "lessons learned":

Lesson 1. Develop one standardized program.
Lesson 2. Develop a scientifically-rooted and evidence-based program.
Lesson 3. Incorporate what we know from the science of learning: engage trainees with interactive learning.
Lesson 4. Focus on sustaining the team behaviors on the job.
Lesson 5. Conduct evaluation at all levels of the Kirkpatrick Hierarchy. (Alonso et al., 2006: 404–405; emphasis in orginal)

Conclusion

Team effectiveness is a product of teamwork (teamwork + interpersonal phases), taskwork (transition + action phases), performance (quality, quantity, time), and the value delivered to the customer (customer satisfaction + team member satisfaction). Team effectiveness is represented by the following formulae:

$$TE = (TW + IP) + TK(TP + AP) + PF + CV \text{ (Turner et al., 2020)}$$

Of all the components of team effectiveness, teamwork skills have been shown to have the greatest impact. Organizations that train and develop strong teamwork skills will have an advantage over competitors that have not developed teamwork skills. High-performance teams require strong teamwork skills, and without these competencies, teams will remain average and organizations could fall behind in complex and disruptive environments.

We believe that the pace of technological innovation plus rapid changes in the global economy, combined with huge demographic shifts now under way, will raise the stakes for team performance. Average will die. High mediocracy won't be enough to win and sustain success. (Karlgaard and Malone, 2015: xi)

Training initiatives need to identify the specific teamwork skills and competencies that are essential to the organization's contextual setting. Training should be designed to deliver deep learning in support of these teamwork skills. Training should implement strong briefing and debriefing exercises to aid team learning, to support development of shared mental models, and to provide teams with the tools to be adaptive.

Example!!

When contrasting professional sports teams and special forces teams in the military with those in organizations, one defining factor stands out. Professional sports teams and special forces constantly train and practice as a team. Organizations spend minimal time allowing their teams to practice, thus preventing teamwork skills from developing and taking root. Team must be able to develop the requisite teamwork skills specific to the organization's contextual setting as much as, if not more than, developing skills to complete required tasks. A stronger focus needs to be placed on team effectiveness and on each of its components, especially on teamwork.

Just as teamwork skills must be developed before any team can be a high-performing team, the same is true for MTSs. Although the components for team effectiveness that were presented in this chapter apply to most individual teams, they become more complicated when you place teams in an MTS. Teamwork is slightly more involved, and teams must be able to operate as a holistic self-organizing unit. MTSs provide more variables to the team effectiveness formulae. MTSs and team effectiveness for MTSs is presented in the next chapter.

References

Alonso A, Baker DP, Holtzman A, et al. (2006) Reducing medical error in the military health system: How can team training help? *Human Resource Management Review* 16: 396–415.

Alonso A and D'unleavy DMA-G, Norma. (2013) Building teamwork skills in healthcare: The case for communication and coordination competencies. In: Salas E and Frush K (eds) *Improving patient safety through teamwork and team training.* New York, NY: Oxford University Press, 41–58.

Behfar KJ, Mannix EA, S. PR, et al. (2010) Conflict in small groups: The meaning and consequences of process conflict. *Small Group Research* 42: 127–176.

Christodoulou I, Babalis D, and Gymnopoulos. (2008) Building teamwork in hospitals: Detecting the situation in an example from Greece. *International Journal of Health Science* 1: 74–77.

Devine DJ. (2002) A review and integration of classification systems relevant to teams in organizations. *Group Dynamics: Theory, Research, and Practice* 6: 291–310.

Dihn JV and Salas E. (2017) Factors that influecne teamwork. In: Salas E, Rico R, and Passmore J (eds) *The Wiley Blackwell handbook of the psychology of team working and collaborative processes.* Malden, MA: Wiley, 15–41.

Driskell JE, Salas E, and Driskell T. (2018) Foundations of teamwork and collaboration. *American Psychologist* 73: 334–348.

Edmondson A. (1999) Psychological safety and learning behavior in work teams. *Administrative Science Quarterly* 44: 350–383.

Edmondson AC. (2019) *The fearless organization: Creating psychological safety in the workplace for learning, innovation, and growth,* Hoboken, NJ: Wiley.

Forsyth DR. (2014) *Group dynamics.* Belmont, CA: Cengage Learning.

Gilbert TF. (2007) *Human competence: Engineering worthy performance.* San Francisco, CA: Wiley.

Gladstein DL. (1984) Groups in context: A model of task group effectiveness. *Administrative Science Quarterly* 29: 499–517.

Greer LL and Dannals JE. (2017) Conflict in teams. In: Slalas E, Rico R, and Passmore J (eds) *The Wiley Blackwell handbook of the psychology of team working and collaborative processes.* Malden, MA: Wiley Online Library, 317–343.

Guzzo RA and Dickson MW. (1996) Teams in organizations; Recent research on performance and effectiveness. *Annual Review of Psychology* 47: 307–338.

Hackman RJ. (2011) *Collaborative intelligence: Using teams to solve hard problems.* San Francisco, CA: Berrett-Koehler.

Hackman RJ and Morris CG. (1975) Group tasks, group interaction process, and group performance effectiveness: A review and proposed integration. *Advances in Experimental Social Psychology* 8: 45–99.

Ilgen DR, Hollenbeck JR, Johnson M, et al. (2005) Teams in organizations: From input-process-output to IMOI models. *Annual Review of Psychology* 56: 517–543.

Jehn KA and Chatman JA. (2000) The influence of proportional and percetpuatl conflict composition on team performance. *International Journal of Conflict Management* 11: 56–73.

Karlgaard R and Malone MS. (2015) *Team Genius: The new science of high-performing organizations.* New York, NY: Harper Collins.

Kozlowski SWJ, Grand JA, Baard SK, et al. (2015) Teams, teamwork, and team effectiveness: Implications for human system integration. In: Boehm-Davis D, Durso F, and Lee J (eds) *APA handbook of human systems integration.* Washington, DC: APA, 555–571.

Kozlowski SWJ and Ilgen DR. (2006) Enhancing the effectiveness of work groups and teams. *Psychological Science of Work Groups and Teams* 7: 77–124.

Lewis K. (2003) Measuring transactive memory systems in the field: Scale development and validation. *Journal of Applied Psychology* 88: 587–604.

Liang DW, Moreland R, and Argote L. (1995) Group versus individual training and group performance: The mediating role of transactive memory. *Personality and Social Psychology Bulletin* 21: 384–393.

Lichtenstein BB, Uhl-Bien M, Marion R, et al. (2006) Complexity leadership theory: An interactive perspective on leading in complex adaptive systems. *Management Department Faculty Publications* 8.

Lim B-C and Klein KJ. (2006) Team mental models and team performance: A field study of the effects of team mental model similarity and accuracy. *Journal of Organizational Behavior* 27: 403–418.

Magpili NC and Pazos P. (2018) Self-managing team performance: A systematic review of multilevel input factors. *Small Group Research* 49: 3–33.

Marks MA, Mathieu JE, and Zaccaro SJ. (2001) A temporally based framework and taxonomy of team processes. *Academy of Management Review* 26: 356–376.

Mathieu JE, Gallagher PT, Domingo MA, et al. (2019) Embracing complexity: Reviewing the past decade of team effectiveness research. *Annual Review of Organizational Psychology and Organizational Behavior* 6: 17–46.

Mathieu JE, Heffner TS, Goodwin GF, et al. (2005) Scaling the quality of teammates' mental models: Equifinality and normative comparisons. *Journal of Organizational Behavior* 26: 37–56.

Mathieu JE, Maynard TM, Rapp T, et al. (2008) Team effectiveness 1997–2007: A review of recent advancements and a glimpse into the future. *Journal of Management* 34: 410–476.

Mayer RC, Davis JH, and Shoorman DF. (1995) An integrative model of organizational trust. *Academy of Management Review* 20: 709–734.

McGrath JE. (1964) *Social psychology: A brief introduction.* New York, NY: Rinehart and Winston.

Rico R, Sanchez-Manzanares M, Gil F, et al. (2008) Team implicit coordination processes: A team knowledge-based approach. *Academy of Management Review* 33: 163–184.

Salas E, Cooke NJ, and Rosen MA. (2008) On teams, teamwork, and team performance: Discoveries and developments. *Journal of the Human Factors and Ergonomics Society* 50: 540–547.

Salas E, Sims DE, and Burke SC. (2005) Is there a "big five" in teamwork? *Small Group Research* 36: 555–599.

Salas E, Tannenbaum SI, Kozlowski SWJ, et al. (2015) Teams in space exploration: A new frontier for the science of team effectiveness. *Current Directions in Psychological Science* 24: 200–207.

Smith-Jentsch KA, Cannon-Bowers JA, Tannenbaum SI, et al. (2008) Guided team self-correction: Impacts on team mental models, processes, and effectiveness. *Small Group Research* 39: 303–327.

Tannenbaum SI and Cerasoll CP. (2012) Do team and individual debriefs enhance performance? A meta-analysis. *Human Factors* 55: 231–245.

Turner JR. (2016) Team cognition conflict: A conceptual review identifying cognition conflict as a new team conflict construct. *Performance Improvement Quarertly* 29: 145–167.

Turner JR and Baker R. (2017) Team emergence leadership development and evaluation: A theoretical model using complexity theory. *Journal of Information and Knowledge Management* 16: 17.

Turner JR, Baker R, Ali Z, and Thurlow N (2020) A new multiteam system (MTS) effectiveness model. *Systems* 8(2), 12.

Weaver SJ, Feitosa J, Salas E, et al. (2013) The theoretical drivers and models of team performance and effectiveness for patient safety. In: Salas E and Frush K (eds) *Improving patient safety through teamwork and team training.* New York, NY: Oxford University Press, 3–26.

Yasmi Y, Schanz H, and Salim A. (2006) Manifestation of conflict escalation in natural resource management. *Environmental Science and Policy* 9: 538–546.

CHAPTER 12

Multiteam Systems: Scaling

Multiteam Systems

After shifting to a team-based organizational structure, the biggest problem comes in managing and organizing teams of teams, or scrum of scrums, or what is called in the literature multiteam systems (MTS). MTSs are defined as follows:

> Two or more teams that interface directly and interdependently in response to environmental contingencies toward the accomplishment of collective goals. MTS boundaries are defined by virtue of the fact that all teams within the system, while pursuing different proximal goals, share at least one common distal goal; and in doing so exhibit input, process and outcome interdependence with at least one or other teams in the system. (Mathieu et al., 2001: 290)

In considering this definition, some key defining features need to be discussed further. First, an MTS is structurally formed when two or more teams are working toward a common goal. This brings us to the second feature of MTSs—that is, goals. Each team within an MTS, we will call them *component teams*, has its own set of team goals. These are called proximal goals.[1] Component teams that are working toward shared multiteam goals are working toward the MTS goal, also called the component team's distal goal. Third, MTSs are structured around component teams having similar distal goals.[2] Component teams can have multiple, and different, distal goals from other component teams within the same MTS. Component teams within the same MTS, however, will share at least one common distal goal that connects them to the overall MTS.

---

1 Proximal refers to near. In the context of MTS, it applies to the team.
2 Distal means distant. In the context of MTS, distal refers to the higher level or the MTS.

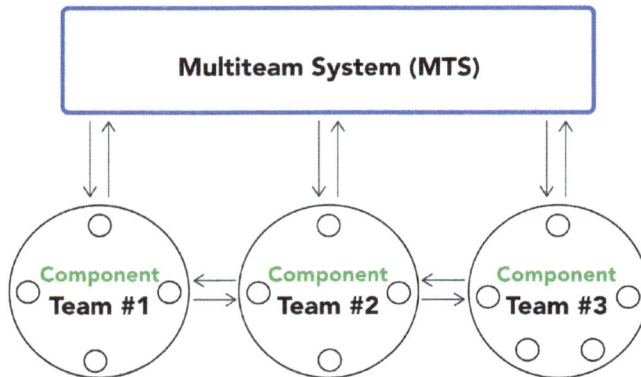

FIGURE 12.1. The MTS

Much in the same manner that team members, within a team, work both independently and interdependently toward that team's goal, MTSs operate in a similar manner. MTSs are composed of component teams working independently and interdependently toward goal accomplishment. This goal accomplishment is the team's distal goal that is connected to the MTS's goal or objective, also identified as the superordinate goal.[3] Component teams have autonomy to identify and design the required proximal goals required to accomplish their distal goal. Component teams work independently from other component teams, and interdependently with other component teams, as needed to accomplish both proximal and distal goals.

Figure 12.1 shows a basic diagram of a simple MTS composed of three component teams. Each component team is working toward its own proximal goals and also is aligned with the MTS through its shared distal goal with the MTS and with other component teams. The distal goals are represented by the arrows in Figure 12.1. Not every component team needs to have the same distal goal as the other component teams, but their goals do contribute to the superordinate goal of the MTS. The arrows in Figure 12.1 also represent the independence of each component team. Although each component team works independently toward its own proximal goals, each one also work interdependently with the other component teams and with the MTS to achieve distal goals.

Levels of Analysis

The processes between a team and an MTS are similar but at different levels of analysis. For example, when looking at teams, we are talking about individual team members working individually (independently) as well as working with

---

3   A superordinate goal identifies the primary goal of an MTS.

other team members (interdependently). Teams are composed of two levels of analysis: individual and team. In this perspective, individuals make up the lower level of analysis and the team represents the higher level of analysis. In contrast, when viewing MTS, we are adding a new level of analysis, totaling to three levels of analysis: individual, team, and MTS. Here, the MTS is the higher level of analysis and the component teams and individuals make up the lower levels of analysis.

Figure 12.1 shows that each component team consists of individual team members (represented by the small circles inside each component team). At the team level, the lower level of analysis is at the individual team member level within each component team. The higher level of analysis is at the component team level. For example, the individual team members within Component Team 1 (four in total; see Figure 12.1) are affected by the team's goals. This represents a higher-level unit (team) influencing a lower-level unit (team member). Likewise, the team is affected by the individual team members. This represents a lower-level unit (team member) influencing a higher-level unit (team). When looking at MTSs, the higher-level unit would be the MTS, which influences each component team (lower level) within the MTS. This is an example of a higher-level unit influencing a lower-level unit (team). In contrast, each component team can also affect the MTS. This is an example of a lower-level unit (component team) influencing a higher-level unit (the MTS).

Goal Hierarchies

In describing the hierarchies of goals for MTS, goals are divided into two hierarchical levels: one at the lower levels of analysis (team) and a second at the higher levels of analysis (MTS). The following features of these goal hierarchies are considered essential toward effective MTS:

> (a) MTS goal hierarchies have a minimum of two levels; (b) goals at higher levels entail greater interdependent actions among more component teams than goals at lower levels; (c) the superordinate goal at the apex of the hierarchy rests on the accomplishment by component teams of all lower order goals; (d) higher order goals are likely to have a longer time horizon than lower order goals; and (e) goals vary in their priority and valence. (Zaccaro et al., 2012: 9; see also Mathieu et al., 2001)

MTS Attributes

MTSs can be categorized according to the three types of attributes: compositional, linkage, and developmental (Shuffler et al., 2015; Turner et al., 2019; Zaccaro et al., 2012). Compositional attributes are similar to team composition; the only difference is that the compositional attributes of an MTS involve multiple teams that are designed to work together and coordinate activities to meet the superordinate goal of the MTS and the organization. Compositional attributes for MTSs include the number of component teams, the size of each component team (not all component teams need to be the same size), within organization and cross-organizational communication functions, location of component teams in relation to one another (virtual, colocated), and each component team's motive structure (Shuffler et al., 2015; Turner et al., 2019; Zaccaro et al., 2012).

Linkage attributes are associated to the mechanisms used to connect component teams to other component teams within the same MTS, component teams to the MTS, and the MTS to the organization. Linkage attributes include the communication and coordination activities, the hierarchical architecture of the MTS, the power distance components of the system, and the networking capabilities of the system at each level (component team, MTS, organization; Shuffler et al., 2015; Turner et al., 2019; Zaccaro et al., 2012).

Developmental attributes identify how the MTS evolves over time. During the first inception of an MTS, team members may feel uncomfortable because of the different team and role structures. Developmental attributes can ease this initial transition as well as support the MTS as it grows and matures over time. Examples of developmental attributes include the MTS's initial structure, tenure of the system, direction of development, and development at each stage in the MTS's duration (mapping development and transformations over time; (Shuffler et al., 2015; Turner et al., 2019; Zaccaro et al., 2012).

Once an MTS is put into operation, it must have the opportunity to allow the MTS to work. Each component team, and its members, must work through a learning curve and identify the best techniques that work for them and the given contextual setting. No two MTSs will be the same. Identifying these techniques will produce the main benefits of the MTS: "its responsiveness and adaptability to challenging performance environments" (Uitdewilligen and Waller, 2012: 365). Implementing an MTS and providing the resources necessary to take root can be supported only through leadership. The next section briefly highlights leadership in an MTS.

Leadership

The more component teams that are added to an MTS, the more structural and functional complexity is added to the MTS. This requires additional coordination, and leadership, between component teams to ensure that the component team outcomes align with the expectations of the MTS.

Research has shown that specific leadership functions are required to the success of an MTS. Leaders must be capable of providing strategy and coordination within teams, across teams, and with external stakeholders and customers (DeChurch et al., 2011). Leadership must be capable of focusing on team actions, the collective, as opposed to focusing on the actions of individual team members (DeChurch and Marks, 2006). While leadership coordinates activities between component teams, each component team must manage its own interactions through shared leadership. These two models of leadership provide a hybrid leadership model for functioning in an MTS. From our research, we have identified two types of leadership styles that work well with an MTS: boundary spanners and shared leadership. Descriptions of both, from our research, as they relate to MTS are provided next.

Boundary Spanners

In leading military units, leaders need to monitor and keep up with activities, interact within and between teams, and remain adaptive and responsive to all stakeholders (Connaughton et al., 2011). Some military departments have utilized boundary spanners as leaders. Boundary spanners (noun) are defined as "organizational members whose primary role is to communicate extensively with constituencies in a relevant environment" (Connaughton et al., 2011: 516). Boundary spanning (verb) is defined as follows:

> Boundary spanning is a concept that encompasses a wide variety of activities, located at the interface between organizational units both within and across formal (e.g., legal) boundaries, from simple information exchange to complex and real-time behavior integrations and coordination. (Davison and Hollenbeck, 2012: 323)

Boundary spanners are situated between the component teams and the organization. Boundary spanners represent the MTS and the component teams that make up the MTS. Boundary spanners identify the goals for the MTS through communications with the executive board, while working on developing proximal goals with each of the component team members. Boundary spanners provide the resources that each component team requires to achieve their goals, while also acting as the single point of contact between component teams and with the MTS and organization.

As a midlevel leadership style, boundary spanners showed promise as one potential style for MTSs (Connaughton et al., 2011). Communicating specialized knowledge and coordinating activities within and between teams became critical for MTSs. Each component team within an MTS had its own proximal goals with at least on distal goal shared with the MTS. Coordinating these activities among each component team to ensure that they accomplished their proximal goals, as well as reassuring that every team's output collectively contributed to the overall MTS's goal, became a complex task that could be performed only outside the component team structure (Turner et al., 2019). Boundary spanning has the following four attributes:

> The type of boundary being spanned, the purpose or function served by a particular boundary-spanning activity, and the degree to which the units on the two sides of the boundary are coupled (i.e., the degree of task interdependence). Additionally, it is important to recognize a fourth attribute, the degree to which a particular boundary-spanning activity is universally applicable across the others, or is more localized to some subset of conditions. (Davison and Hollenbeck, 2012: 326)

For an MTS, the boundary spanner is responsible for the boundary between component teams and the boundaries between each component team and the MTS. In Figure 12.1 these boundaries are represented as the double arrows, identifying the boundaries for the boundary spanner. Not shown in Figure 12.1 is the boundary at which the boundary spanner communicates with the organization and customer. Figure 12.1 is a basic drawing of an MTS, but it provides an idea of what a boundary spanner leadership structure might look like. The exact structure of a boundary spanner will vary from one organization to the next. The concept presented in Figure 12.1 shows one boundary spanner per component team. Ideally, it would be more efficient to have only one boundary spanner per MTS. This will work only if the boundary spanner is in this role as a full-time leader with no other job responsibilities. If this is a part-time role, then one boundary spanner would make more sense.

The purpose of the boundary spanner will vary, but at a minimum, the boundary spanner should identify the distal goals for the component teams and work with each component team to identify their proximal goals. The boundary spanner will provide resources and foster interactions between component teams and coordinate activities with the organization. The level of the interactions between component teams that will be required will depend on the type of coupling necessary between component teams. This means that more interactions will be fostered and managed for closely coupled tasks (tasks requiring component teams to work interdependently) and fewer interactions will be fostered for loosely coupled tasks (mostly independent component team tasks with minimal interdependency between component teams).

The fourth boundary-spanning attribute highlights that some boundary-spanning activities may require more interactions at the organizational level and even at the customer level. Other activities could require fostering activities external of the organization (e.g., industry level, global partnerships). Other boundary-spanning activities may need to be contained only within the MTS. The point is that the level of interaction required by any boundary spanner will vary from one organizational setting to the next.

Shared Leadership

In existing MTS models, the leadership functions follow a shared leadership model rather than a traditional leader-follower paradigm: "The leadership role in the [MTS] . . . does not easily fit a leader-follower paradigm where decision authority stays with the leader. It is assumed that a shared leadership model plus a facilitating, impartial leader (coordinator) will be more effective" (Goodwin et al., 2012: 64). Shared leadership is defined as follows:

> A dynamic, interactive influence process among individuals in groups for which the objective is to lead one another to the achievement of group or organizational goals or both. This influence often involved peer, or lateral, influence and at other times involves upward or downward hierarchical influences. (Pearce et al., 2007: 282; see also Bienefeld and Grote, 2014)

In discussing which task characteristics are called for in shared leadership, Pearce highlighted tasks that were highly interdependent, which required a high level of creativity, and tasks that involved high levels of complexity, placing shared leadership idea for team-based systems (Pearce, 2004). Shared leadership should be considered whenever the three characteristics of interdependence, creativity, and complexity are in play (Pearce, 2004). Even with shared leadership, there is still a need of some level of traditional leadership to support the team's efforts. Shared leadership was best supported through traditional leadership styles: "directive, transactional, transformational and empowering" (Pearce, 2004: 53). Support for the team came from a leader who clarified the organization's goal and the team's role in achieving this goal (Pearce, 2004). This support needed to be in such a manner that the team members did not feel they were being managed, leadership must not have exerted power or influence over individual team members. In the worlds of Pearce, "shared leadership support should be inherently cautious" (Pearce, 2004: 54).

Shared leadership also looked at the shared aspect of leadership, identifying all leadership as being shared, at one time or another (Pearce et al., 2014). Shared leadership involved "the serial emergence of both official and unofficial leaders as part of a simultaneous, ongoing, mutual influence process" (Pearce et al., 2014: 276). From the perspective of shared leadership, leadership involved multiple players at every level. The amount and method in which this sharedness took place varied. The role of being the leader not only could have been based on power-distance, knowledge, or experience but also could have been designed around a coaching or mentoring structure. This places trust as a critical aspect to successful shared leadership (Pearce et al., 2014). Building a transactive memory system is also necessary, providing team members with information relating to, who knew what, and who had what knowledge and experience (Pearce et al., 2014).

Other research has highlighted some form of shared leadership for MTSs (Shuffler et al., 2015) in conjunction with some type of oversight. Either from an individual leader or by a leadership team perspective, shared leadership and traditional vertical leadership models should "work in tandem" (Pearce et al., 2014: 285). The specific leadership style did vary depending on the size and structure of the MTS. Leadership was also different for MTSs that were housed within one corporation or entity. Generally, researchers have found that "sharing of leadership among team members can complement the formal leadership structure" (Shuffler et al., 2015: 679). Exactly what this leadership model looked like was different for each MTS.

Hybrid Leadership Style

As highlighted previously, MTSs have a variety of leadership structures. The point is to find which leadership style works best for one's contextual setting. In the previous chapters on distributed leadership (Part III) in this book, we discussed the need of incorporating a blended style of leadership for team-based systems.

New Organizational Structure
(Required for Team-Based Systems)

Executive Level

BLENDED   **STRUCTURE**

Team-Based System

SHARED LEADERSHIP

Leadership

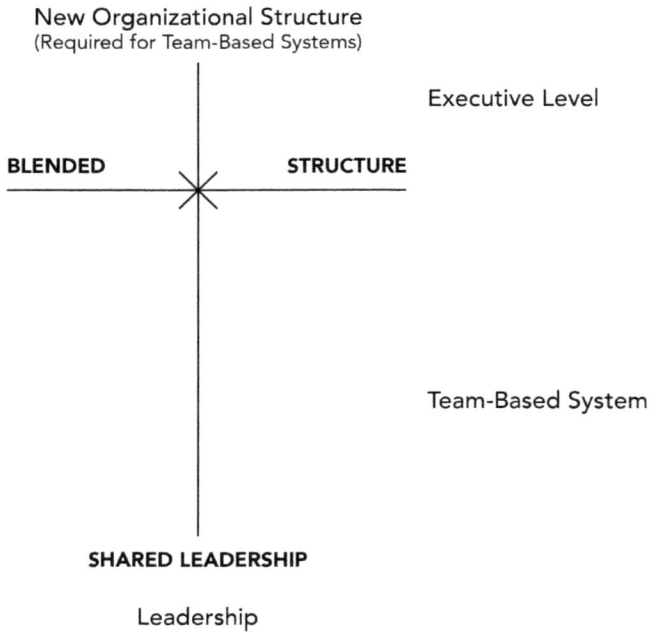

FIGURE 12.2.   Blended Leadership Model for Team-Based Systems

Figure 8.1 was introduced in Chapter 8, Team and Distributed Leadership, and is duplicated here as Figure 12.2. The shared leadership model would be the preferred leadership style for the component teams in an MTS. Leadership at the executive level would be more favorable as discussed in Chapter 9, Strategic, Instrumental, and Global Leadership. Connecting the two leadership styles, between the component teams and the executive-level leadership structure, is the boundary spanner and the MTS. Figure 12.3 provides a general preview of what a basic MTS looks like, including its goal structures and leadership structures.

Figure 12.3 incorporates the distributed leadership concept surrounding an MTS. The leadership style at the component team level is shared leadership, with boundary spanners included at the MTS level, and the hybrid leadership styles (strategic, instrumental, global leadership) at the executive and organizational levels. The proximal goals are shown at the team level, the distal goals at the MTS level, and organizational goals at the executive and organizational level.

Multiteam System Effectiveness

In the previous chapter, we discussed the different team effectiveness framework models, the primary model of these is the input-process-output (IPO) framework

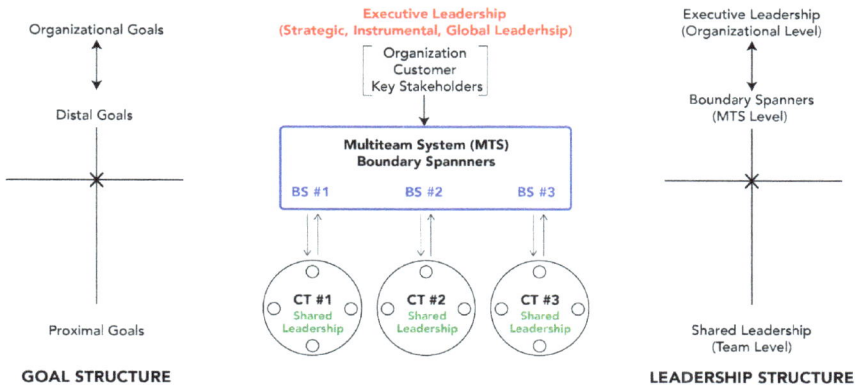

FIGURE 12.3.   MTS Leadership Structure

(McGrath, 1964). As previously discussed, interdependencies for MTSs operate slightly differently from those of a single team. Using the IPO framework, these interdependencies have been categorized as input interdependencies, process interdependencies, and outcome interdependencies (Zaccaro et al., 2012). As additional component teams are added to an MTS, interteam action processes, those interactions required between component teams, become more essential to MTS effectiveness (Zaccaro et al., 2012).

Input interdependencies refer to the sharing by component teams, including sharing of human resources, information (creation, transfer, storage), technology and materials (resources), and capital (Zaccaro et al., 2012). Process interdependencies are associated with the interactions required between the component teams, including "boundary spanning and environmental sensemaking, task ordering and tactical planning, communicating key information, the timing and coordination of sequential and synchronous actions, and the monitoring and backup of MTS actions" (Zaccaro et al., 2012: 10; see also Marks et al., 2001). Consideration must be made to sequential and parallel, or reciprocal, process interdependencies (Zaccaro et al., 2012). Sequential processes involve tasks in which one component team depends on another component team finishing their task before being able to complete its task. Parallel processes involve more than one component team working on the same task at the same time.

Outcomes interdependencies relate to the outcomes of each component team as they relate to MTS goal accomplishment (Zaccaro et al., 2012). In the same manner that a single team's goal must be connected to the organization's goals, each component team's outcomes must be connected, in one way or another, to some part of the superordinate goal, that of the MTS. The aggregate of all component team outcomes will result, over time, in accomplishing the superordinate goal of the MTS and, ultimately, of the organization. Complications arise when component team outcomes are not related to the distal goal, or when component team outcomes are incorrect and not applicable toward goal achievement of the MTS.

The effectiveness of an MTS become more complicated compared with the effectiveness of an individual team. The formulae for team effectiveness for an individual team was provided in Chapter 11, Team Effectiveness. The formulae is as follows:

$$TE = (TW + IP) + TK(TP + AP) + PF + CV \text{ (Turner et al., 2020)}$$

Team effectiveness (TE) is a function of teamwork (TW) and interpersonal processes (IP) plus taskwork (consisting of transition phases, TP, and action phases, AP), plus performance (PF; quality, quantity, time), plus customer value (CV; feedback, customer satisfaction, team member commitment and satisfaction). Given that an MTS is a different level of analysis compared with the component team, we have two formulae for MTSs. One at the team-level unit (lower level of analysis), and one at the MTS-level unit (higher level of analysis).

Team effectiveness at the MTS level requires interactions between component teams as well as interactions between each component team and the boundary spanner (the MTS). Multiteam effectiveness would be equal to component team to component team interactions (C-C) and component team to boundary spanner or MTS interactions (C-BS). The performance required of the MTS (PFmts) and the customer value is more directly related to the actual customer (CVmts). The team member satisfaction and commitment components are at the team level and not at the MTS level.

Team effectiveness at the MTS level (MTS TE) produces the following formulae:

MTS TE (higher level) = C-C + C-BS + Pmts + CVmts + SUM[TE (lower level)]
SUM[TE (lower level)] = Component Team 1 [(TW + IP) + TK(TP + AP) + PF + CV] + ... + Component Team n [(TW + IP) + TK(TP + AP) + PF + CV]. (Turner et al., 2020)

The sum of team effectiveness at the lower level results in the aggregate of team effectiveness for each component team. For example, if there were five component teams, then this will represent the team effectiveness for five component teams, a summation. These formulae are not being presented to represent a continuous variable or a specific acceptable range for a numerical value. These formulae, however, are broken down into the components that result in team effectiveness at both the component team and the MTS levels of analysis. Managing each of these components, and designing training surrounding each of these components, are recommended to achieve maximum effectiveness at the MTS level.

Using MTS to Solve Complex Problems

> The field of social ecology has been described as a transdisciplinary field organized around certain *descriptive, explanatory,* and *transformative* goals . . . *to describe the structure of human environments,* especially their *natural, built, social,* and *virtual features,* and the processes by which these different parts of our surroundings change over time and influence each other. (Stokols, 2018: 63; emphasis added)

Structural human environments are represented by a variety of environments at varying scales (Figure 12.4), including natural environment, built environment, sociocultural environment, and virtual environments (Stokols, 2018). Natural environments include nature with all its glory (plants, animals, insects), whereas built environments include the physical environments designed and built by humans. Sociocultural environments include organizational and institutional systems, the legal system along with ethics and norms, including the activities that individuals participate in as a group or community member. Virtual environments include computing and mobility technologies, including the World Wide Web, the Internet of Things, social media, and other digital technologies (Stokols, 2018).

Each of these structural human environments need to be considered when defining a problem. It must be determined which part, if any, of each type of environment are affected by the problem. The alternate is also true: one or more components of a structural environment could be either causing the problem or mediating the effects of the problem. One challenge is to identify the few contextual variables most crucial for understanding the problem: "how we describe and analyze social problems influences our decisions to focus on some rather than

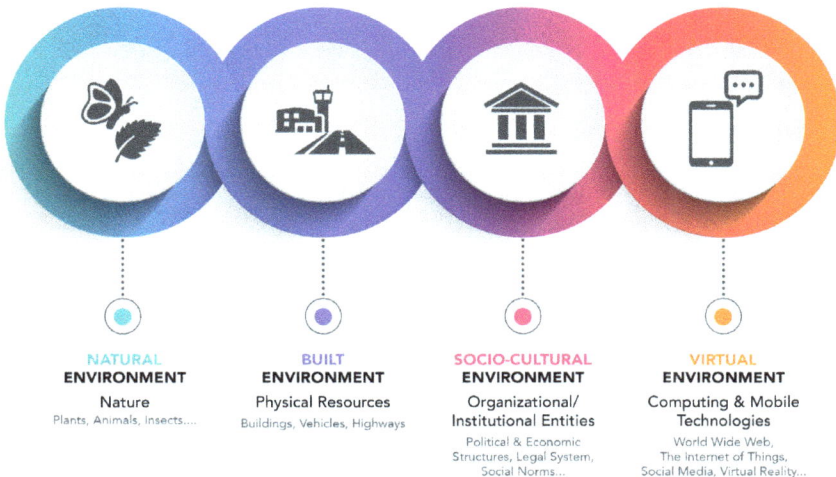

NATURAL ENVIRONMENT	BUILT ENVIRONMENT	SOCIO-CULTURAL ENVIRONMENT	VIRTUAL ENVIRONMENT
Nature	Physical Resources	Organizational/ Institutional Entities	Computing & Mobile Technologies
Plants, Animals, Insects....	Buildings, Vehicles, Highways	Political & Economic Structures, Legal System, Social Norms...	World Wide Web, The Internet of Things, Social Media, Virtual Reality...

FIGURE 12.4. Structural Human Environments: Social Ecology

others, and the chances that our efforts to alleviate them will be successful" (Stokols, 2018: 190).

One method of breaking down complex problems, from the field of social ecology, includes breaking down complex problems into multiple levels, spanning "micro to macro scales of the environment and includes individual, small group, organizational, institutional, community, and global levels of analysis" (Stokols, 2018: xxiii–xxiv). Figure 12.5 provides a general preview of the multiple levels involved in solving some types of organizational complex problems. This multilevel approach also includes multiple stakeholders: politicians, academics, practitioners, and lay people or community members. Analyses also require multilevel analyses techniques over long periods of time (longitudinal) as opposed to single-scale and single-time analysis techniques (Stokols, 2018).

The benefit of expanding the analysis to a multilevel analysis, involving stakeholders from each level of analysis, is that this method provides a better chance of capturing, and confirming, the changing casual mechanisms to a complex problem. By including multiple levels of analysis, smaller problems can be identified at different levels. If a problem was identified at a lower level of analysis at one point in time, and later found not to be present, results would be confusing. The story changes, however, if this same problem also was

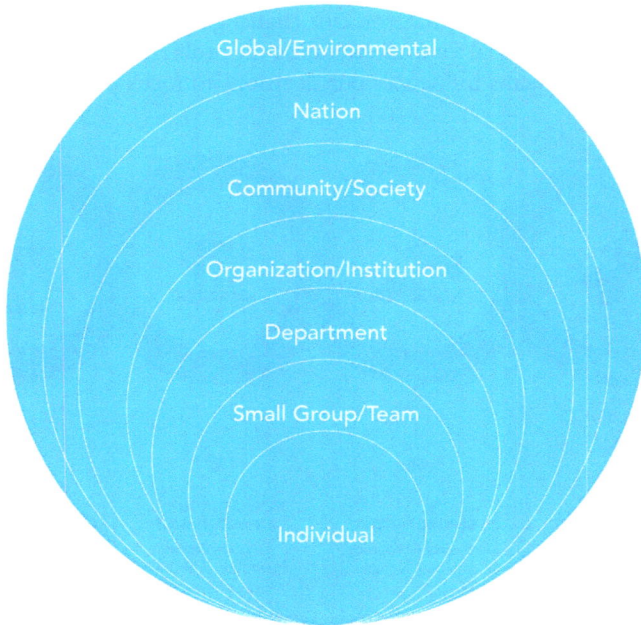

FIGURE 12.5.   Multilevel Approach to Solving Complex Problems

occurring at other levels of analysis. This method could also provide closer inspection of influences that higher level effects may have on lower levels of analysis, and vice versa.

When we span outward, looking at multiple levels, we are able to provide a larger picture surrounding the problem. From a social ecology perspective, analyses are guided by the following core principles:

- Human environments consist of multiple dimensions including natural, built, sociocultural, and virtual (cyber-based) features, some that are directly observable and others subjectively perceived
- People's transactions with their surroundings occur at multiple levels and are nested within bounded environmental contexts (such as homes, workplaces, communities, regions, nations) that are interconnected across varying geographic, social, and temporal scales
- Environments and their inhabitants are dynamic systems where individuals and groups react to changes in their surroundings and, in turn, actively modify the environment to better suit their needs
- Social ecology is inherently transdisciplinary in its approach to understanding people's relationships with their surroundings. It draws on concepts, theories, and methods from several fields and emphasizes an action research orientation by integrating academic and nonacademic perspectives to more effectively analyze and manage complex societal problems. (Stokols, 2018: 67)

Drawing on such a vast collection of data and information could lead to information overload. It is necessary, however, to identify (using theory, empirical evidence, practice, community input) the contextual variables most essential to the phenomenon being studied: "investigators should focus on identifying the *effective context*—that *subset of high-leverage situational variables most crucial for understanding a particular phenomenon*" (Stokols, 2018: 68).

Mapping tools and techniques can help identify relevant variables within a person or groups surrounding, which is known as the contextual mapping phase. Selecting specific, and relevant, situational variables over those that appear to be irrelevant is the next step, which is known as the contextual specification phase. Combining these two phases, contextual mapping and specification phases, into a single theory produces a contextual theory: "those that *predict cross-situational variations in the target phenomenon*" (Stokols, 2018, p. 69).

Structuring a Multiteam System Around a Complex Problem

Utilizing the benefits afforded by an MTS could be one tool to implement the multilevel type of analysis required to solve complex problems. Taking the general configuration of an MTS (Figure 12.1), we can incorporate each of the multilevel elements into its own component team, all bounded within a single MTS.

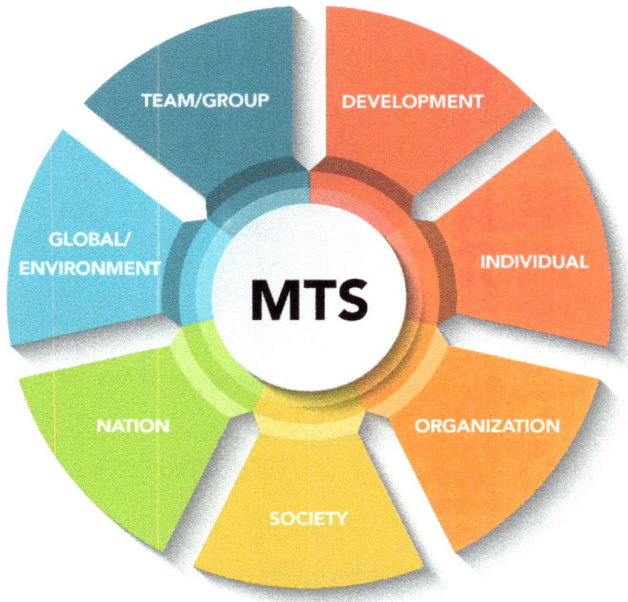

FIGURE 12.6.   Multiteam Structure for Solving Complex Problems

Figure 12.6 outlines what this model could look like. Each component team looks at the problems surrounding its assigned levels of analysis that involves all relevant stakeholders. The individual component team, for example, looks at the impacts that the problem has on individuals in the workplace (for an organizational problem). This analysis involves each of the four structural human environments (natural, build, sociocultural, virtual) and breaks down the problem from the perspective that the problem exists in a complex adaptive system. The characteristics of a complex adaptive system as they relate to the problem also must be defined: path dependency, system history, nonlinearity, emergence, irreducibility, adaptability, operation between order and chaos, self-organization (see Chapter 6, Systems Thinking Versus Complexity Thinking). The same will be done with the other component teams (small group/team, department, organization/institution, community/society, nation, global/environmental) that make up the MTS shown in Figure 12.6. Depending on the type of problem, however, not all structural human environments need to be included in the problem-solving stages. This needs to be determined during the initial phases when designing the MTS.

MTSs are composed of many interconnecting elements. What is not shown in Figure 12.6 are the interactions required between each component team and the interaction, communication, and coordination activities performed by the boundary spanner aimed toward the superordinate goal, that is, solving the complex problem.

Conclusion

When transitioning to a team-based organizational structures, MTSs need to be organized around specific products or projects. How each MTS is designed, including the goals and leadership structure, will be essential to the individual MTS's success. When looking at MTS effectiveness, the components highlighted in Chapter 11, Team Effectiveness and given in Table 11.2, MTS Effectiveness Model, should be considered and developed to achieve MTS and organizational objectives. Embedding MTSs into an organizational structure requires an organizational transformation that includes each of the levels of analysis (individual, team, MTS, and organizational), all levels of goal structures (proximal, distal, organizational), and leadership styles (shared leadership; boundary spanners; strategic, instrumental, and global leadership). Blindly jumping into an MTS, or some version of scaling teams (e.g., team of teams, scrum of scrums), without first putting into place the appropriate and necessary structures (Figure 12.2), will lead to complications and more than likely will result in poor outcomes. Training, development of team members and leaders, and changes in organizational structure to accommodate the MTS is necessary before implementing any type of MTS.

Functional and structural complexity increases when organizations expand to multiple MTSs. The goal and leadership structures are scaled, adding new areas for constraints and bottlenecks to be introduced into the system. Cautionary measures need to be made to *not* add too many component teams to any one MTS. Just as small teams provide the best benefit to solving creativity and complex problems, small MTSs can provide the same advantages to complex and wicked problems (Figure 12.6). Having MTSs that are bogged down with too many component teams causes additional communication and coordination difficulties. Having smaller MTSs would eliminate many of these problems while producing more effective and efficient outcomes.

When implementing multiple MTSs, the structure will vary from one organization to the next. The purpose of this chapter is not to present a single model for organizations to implement, as there is no one universal model. This chapter, however, offers the essential elements that need to be in place to implement successful MTSs in an organization and to solve complex problems. This chapter describes MTSs and explains how they are typically structured, how they are managed, and what their goal structures look like. Each organization, each contextual setting, and each type of project will need to determine exactly how this should be accomplished.

References

Bienefeld N and Grote G. (2014) Shared leadership in multiteam systems: How cockpit and cabin crews lead each other to safety. *Human Factors* 56: 270–286.

Connaughton S, Shuffler M, and Goodwin GF. (2011) Leading distributed teams: The communicative constitution of leadership. *Military Psychology* 23: 502–527.

Davison RB and Hollenbeck JR. (2012) Boundary spanning in the domain of multiteam systems. In: Zaccaro SJ, Marks MA, and DeChurch LA (eds) *Multiteam systems: An organization form for dynamic and complex environments*. New York, NY: Routledge, 323–362.

DeChurch LA, Burke CS, Shuffler ML, et al. (2011) A historiometric analysis of leadership in mission critical multiteam environments. *The Leadership Quarterly* 22: 152–169.

DeChurch LA and Marks MA. (2006) Leadership in multiteam systems. *Journal of Applied Psychology* 91: 311–329.

Goodwin GF, Essens PJMD, and Smith D. (2012) Multiteam systems in the public sector. In: Zaccaro SJ, Marks MA, and Dechurch LA (eds) *Multiteam systems: An organization form for dynamic and complex environments*. New York, NY: Routledge, 53–78.

Marks MA, Mathieu JE, and Zaccaro SJ. (2001) A temporally based framework and taxonomy of team processes. *Academy of Management Review* 26: 356–376.

Mathieu JE, Marks MA, and Zaccaro SJ. (2001) Multiteam systems. In: Anderson N, Ones DS, Sinangil HK, et al. (eds) *Handbook of industrial, work and organizational psychology*. Thousand Oaks, CA: Sage, 289–313.

McGrath JE. (1964) *Social psychology: A brief introduction*. New York, NY: Rinehart and Winston.

Pearce CL. (2004) The future of leadership: Combining vertical and shared leadership to transform knowledge work. *Academy of Management Executive* 18: 47–57.

Pearce CL, Conger JA, and Locke EA. (2007) Shared leadership theory. *The Leadership Quarterly* 18: 281–288.

Pearce CL, Wassenaar CL, and Manz CC. (2014) Is shared leadership the key to responsible leadership? *Academy of Management Perspectives* 28: 275–288.

Shuffler ML, Jimenez–Rodriguez M, and Kramer WS. (2015) The science of multiteam systems: A review and future research agenda. *Small Group Research* 46: 659–699.

Stokols D. (2018) *Social ecology in the digital age: Solving complex problems in a globalized world*. San Diego, CA: Academic Press.

Turner JR, Baker R, Ali Z, and Thurlow, N (2020). A new multiteam system (MTS) effectiveness model. *Systems* 8(2): 12.

Turner JR, Thurlow N, Baker R, et al. (2019) Multiteam systems in an agile environment: a realist systematic review. *Journal of Manufacturing Technology Management* 30: 748–771.

Uitdewilligen S and Waller MJ. (2012) Adaptation in multiteam systems: The role of temporal semistructures. In: Zaccaro SJ, Marks MA, and DeChurch LA (eds) *Multiteam systems: An organization form for dynamic and complex environments*. New York, NY: Routledge, 365–394.

Zaccaro SJ, Marks MA, and Dechurch LA. (2012) Multiteam systems: An introduction. In: Zaccaro SJ, Marks MA, and Dechurch LA (eds) *Multiteam systems: Am organization form for dynamic and complex environments*. New York, NY: Routledge, 3–32.

PART V

Conclusion: Values and Attributes of The Flow System

CHAPTER 13

Conclusion and Resources

Think Differently

R esearching seasoned CEOs who have been identified as being high performing, Zemmel et al. highlighted five themes essential to an organization's success:

> The importance of resetting ambitions to avoid losing momentum; the need to attack silos and fix broken processes; the imperative of rejuvenating leadership talent; the value of building internal and external mechanisms for dissent and disruptive ideas; and the need to deploy leadership capital on bold moves that could help the company succeed over a long horizon. (Zemmel et al., 2018: 100)

Contrast this statement from Zemmel et al. (2018) with the following one that provides an example of why executives are apprehensive to change:

> High-level executives are ambivalent about changing their own behavior. They know perfectly well that their companies need to become more innovative - and they suspect it won't happen unless they're willing to push power, decision making, and resource allocation lower in the organization. But they're terrified that the business will fall into chaos if they loosen the reins. (Ancona et al., 2019: 76)

The former statement provides essential themes that are necessary for executives to continue their organization's success. The latter statement highlights executive's unwillingness, or fear, to release control to achieve the former. To prevent an organization from losing momentum and from reverting back to previously identified failed practices (e.g., Columbia Accident Investigation Board, 2003), it is recommended to continuously reimagine the business using fresh eyes. Examples of this reimagination can be found in Steve Job's story and his successful turnaround of Apple after his initial dismissal from the executive ranks of Apple in 1985. One of Job's talents was in his ability to freely change "his mind when he realized he needed to think differently" (Isaacson, 2014: 94).

The Flow System (TFS) provides a reimagined system for organizations to implement, using complexity thinking to *think differently*, keeping ambitions high through teamwork and autonomous team-based leadership structures. TFS also allows those closest to the customer, team members, to determine the best course of action for both the organization and the customer, as opposed to reverting to the ill-conceived best practices used previously. The team-based hybrid organizational structure presented in TFS not only provides a flatter hierarchical system but also provides a system that is cross-functional throughout an organization, removing the barriers and constraints that often occur from traditional hierarchical organizational structures and silos. TFS also provides an entrepreneurial spirit resulting from the autonomous team-based structures that allow for disruptive strategies to be fostered while also preventing broken processes to take root. The leadership structure of distributed leadership in TFS provides a process that constantly revives leaders throughout an organization, allowing the collective leadership to make bold and disruptive moves across an industry.

TFS is a system that matches the needs highlighted by today's high-performing CEOs. TFS meets the new calls for tomorrows leaders: "Leaders need to look at the organization and the markets in which it plays with fresh eyes and keep evolving their strategy and approach to their team. They can't take their foot off the gas-if anything, they need to push down harder" (Zemmel et al., 2018: 105).

A Transformation

Sometimes things stay the same, even with vast advances in technology and scientific discoveries. For example, in 1945, Lewin highlighted the need to advance managerial practice beyond the manager and the organization: "We are just awakening to the fact that a better knowledge is needed than day by day experience, tradition and memory of an individual or a social group can provide, that we need understanding on a scientific level" (Lewin, 1945: 129). Lewin began applying research to social issues in the workplace, which marked the beginning of social psychology as we know it today. Although we have made advances in the field of social psychology and in other cognate fields of study related to social dynamics, today's call would be for advancing research to include methods and techniques to address complex issues and environments in social settings. One such social setting is the organization. TFS is one such model that any organization or institution can begin applying to address their complex issues and environments.

Realigning an organization's focus to address tomorrow's disruptive and complex environments requires not only new organizational goals and values but also a change or modification to an organization's existing purpose, strategy, and structure. Organizations are tasked with building and modifying their structures according to the organization's desired results (Drucker, 2007). If an organization wants to be adaptive to challenge disruptive forces and to adopt complexity thinking to combat complex and wicked problems that may challenge them in the near future, they must be willing to change strategically and structurally. As Drucker has noted:

> A business should always analyze its organization structure when its strategy changes. Whatever the reason—a change in market or in technology, diversification or new objectives—a change in strategy requires a new analysis of the key activities and an adaptation of the structure to them. Conversely, reorganization that is undertaken without change in strategy is either superfluous or indicates poor organization to begin with. (Drucker, 2007: 171–172)

TFS provides organizations and institutions with the tools and techniques to begin the transformational process to implementing new organizational team-based structures to better meet the demands of tomorrow's disruptive and complex environments. This transformation cannot be achieved by focusing only on one of the helixes of TFS. Each of the three helixes in the Triple Helix of Flow must be addressed: complexity thinking, distributed leadership, and team science. During this transformation, each of the three helixes need to be integrated into all organizational functions to obtain flow. Without this integration, flow cannot be achieved, preventing value to be delivered to the customer. The concept of the Triple Helix of Flow is essential to any organization and must be at the forefront of any organizational transformation.

It's Not Necessarily New, It Just Needs to be Done

The concepts presented in TFS are not necessarily new. They have been around for some time—in research studies, in magazine articles, in training manuals, and in various organizational archives. This knowledge also has been combined with practice, the experiences that each of us (the authors) have gained through practice, including the experiences of those we had worked with over the years. This fusion of knowledge and experience is in alignment with empirical research that dates as far back as Alfred North Whitehead's important quote: "Everything of importance has been said before by somebody who did not discover it" (Stigler, 2010: 287). The key difference, or contribution, that TFS provides is that it incorporates both theory and practice into one comprehensive conceptual–theoretical model and connects three essential constructs or helixes that must be interconnected before achieving flow.

Although many of the concepts presented in TFS may not be new, they have been either ignored or not put into practice. For a number of reasons, certain beneficial practices have not been put into action (e.g., power, lack of knowledge, timing, not knowing). We are not focusing on the reasons why; we are just making the point that it is time to act and to put these practices to action. Otherwise, facing disruptive competitors and complex problems will be nearly impossible in the future.

One example in which similar information already has been presented in the literature can be found in *The Fundamentals of Industrial Management,* Series 1: Civil Communication Section (CSS), by Homer Sarasohn and Charles Protzman (1998).[1] This course is based on theory and their experiences from the rebuilding efforts of Japan during 1946–1950, under the guidance of General Douglas MacArthur, the following was provided when discussing the importance of teamwork to organizations:

Teamwork

It is a natural human trait, possessed by every one of us, to want to excel. We want to be personally prominent, to achieve a position where we will stand out, where we will feel that we are important, that we are necessary. In its most exaggerated form, this trait leads us to believe we are the only one who can properly do the job, and that others with whom we associate are all inferior in brains, ability or experience. Then we become the "indispensable man," without whom no decisions can be made, no ideas worthy of consideration conceived, and no sound action taken.

But along with this universal trait is another that is probably equally as strong. This is the reaction any of us has in dealing with someone who considers himself "indispensable." We resent the omnipotence of his decisions, and chafe because we don't have an opportunity to express our own views, or show what we are capable of doing. And whether this person is a superior, or an associate, we do not work together with him. As a matter of fact we avoid him as long as he continues to take an omnipotent attitude.

However, no company can afford the luxury of permitting individuals to place their own position or progress ahead of the good of the company, because when this happens, every employee who has the ambition or desire to do so will have an example and precedent to act upon his own private goal or objective. Usually, such individual goals are purely selfish, and the result is a number of independent units or groups, each led or dominated by a strong personality, and each going its own separate way without regard to the effect it will have on other employees or on the company. Experience has shown, over and over, that companies in which this condition exists will ultimately deteriorate and fail. This is simply one of the economic facts of life that cannot be avoided or ignored.

When an individual goes to work in a company, his first interest is in making a living. Next, he is interested in bettering his position or making a better living. If he is a workman who is part of a group he soon realizes or is made to realize by his associates that his contribution to the group effort is important because unless the job is done properly he will soon be looking for another job.

---

1   An e-book transcription of the version presented at the 1949 Tokyo seminar was prepared by Nick Fisher and Suzanne Lavery of Value Metrics Australia and is widely available online. The documents, which formed the final English version that was translated by Bunzaemon Inoue and others and published in 1952 in Japanese by Diamond Press, are located in the CSS Archives (Hackettstown, NJ) and the Drucker Institute (Claremont, CA).

But when this individual becomes a supervisor, a part of management, he often fails to realize that he has assumed the obligation and accepted the responsibility of striving for certain objectives of the company. These objectives are to meet the company's obligations for programs, schedules, quality and costs. For the success of the company, and the ultimate achievement of his personal objectives, he must work primarily for the company's objectives. These can best be accomplished through teamwork, which can be defined as follows:

> "Teamwork is the work done by a number of associates, all subordinating personal prominence to the efficiency of the whole."

If we consider this definition for a moment, we will see that teamwork is something that is voluntary. We cannot depend on a law or a rule that says that we must have teamwork, or cooperation. People must want to work together and must realize that by working together they also have the best chance of reaching their own individual objectives.

But where must teamwork start? Earlier we talked about the fact that each of us, either consciously or unconsciously, picks up mannerisms, methods of approach and ways of doing things from those for which we work.

If our superiors are "indispensable men," then we also tend to be indispensable in the same disagreeable and destructive way. On the other hand, if our superiors recognize the advantages and the necessity of working together with their associates and subordinates, then we tend to do the same.

So, for full effectiveness, teamwork and cooperation must be established by the attitude and example of each management level from the president down. It must be encouraged by the day-to-day manner in which every management employee does his work and deals with other people. (Sarasohn and Protzman, 1998: 163–164)

This example touches on both teamwork and leadership. It highlights the point that leadership must be distributed, as there is no room for any employee to think they are indispensable, leadership needs to be collaborative at all levels—distributed leadership. Teamwork also involves collaborative effort at all levels of the organization working together, from the president (CEO) down. This example also highlights the point that achieving organizational goals can, or must, be achieved through teamwork. This early description of teamwork makes the connection that leadership and teamwork are intricately connected; each one relies on the other.

A second example touches on the concept of complexity, related to software development and engineering. This example comes from Frederick Brook's book titled *The Mythical Man-Month: Essays on Software Engineering*, originally published in 1975 with a dedication edition published in 1995. Brooks mentioned teamwork and management, loosely connecting the three helixes in the Triple Helix of Flow—complexity thinking, distributed leadership, and team science:

> A scaling-up of a software entity is not merely a repetition of the same elements in larger size; it is necessarily an increase in the number of different elements. In most cases, the elements interact with each other in some nonlinear fashion, and the complexity of the whole increases much more than linearity.

The complexity of software is an essential property, not an accidental one. Hence descriptions of a software entity that abstract away its complexity often abstract away its essence. Mathematics and the physical sciences made great strides for three centuries by constructing simplified models of complex phenomena, deriving, properties from the models, and verifying those properties experimentally. This worked because the complexities ignored in the models were not the essential properties of the phenomena. It does not work when the complexities are the essence.

Many of the classical problems of developing software products derive from this essential complexity and its nonlinear increases with size. From the complexity comes the difficulty of communication among team members, which leads to product flaws, cost overruns, schedule delays [preventing value to be delivered to the customer]. From complexity comes the difficulty of enumerating, much less understanding, all the possible states of the program, and from that comes the unreliability. From the complexity of the functions comes the difficulty of invoking those functions, which makes programs hard to use. From complexity of structure comes the difficulty of extending programs to new functions without creating side effects. From complexity of structure comes the unvisualized states that constitute security trapdoors.

Not on technical problems but management problems as well come from the complexity. This complexity makes over-view hard, thus impeding conceptual integrity. It makes it hard to find and control all the loose ends. It creates the tremendous learning and understanding burden that makes personnel turnover a disaster. (Brooks, 1995: 183–184)

This sample highlights that agents interact more as complexity increases and that the processes (e.g., interactions, communication, knowledge flow) that take place in complex systems are nonlinear. When solving complex problems, the goal is to identify patterns in complexity rather than to eliminate the elements that cause these patterns. Identifying what barriers to place around open complex adaptive systems is an important practice in complexity thinking. Brooks also hints at the idea that as functional complexity increases, structural complexity also must change to accommodate the new functions of the complex adaptive system. This is relevant in organizations in that they need to change their organizational structures, to team-based structures and to distributed leadership, to accommodate increases in functional complexity. The reference to "management problems" also eludes to what we have described as distributed leadership. Brooks' example indirectly connected each of the three helixes—complexity thinking by highlighting increasing complexity and nonlinear processes, distributed leadership through changes to structural complexity and management problems, and team science by highlighting the importance of communicating to team members in times of complexity.

This last example comes from the complexity literature and highlights the concepts of self-organization, emergence, and adaptability and also touches on each of the three helixes in TFS:

While traditional models of structures focused on change and evolution through "top-down" centralized control, a complexity approach emphasizes the ability of un-coordinated, "bottom-up" dynamics to generate coherent structure. The notion of "structure" is used here to describe the internal mechanism developed by the system to receive, encode, transform and store information, and to react to this information through some form of output. Complexity informs us that internal structure can evolve without the intervention of an external designer or the presence of some centralized form of internal control. If the capacities of the system can satisfy a number of constraints, it can develop a distributed form of internal structure through a process of self-organization. The structure that becomes apparent through this process is neither a passive reflection of the external environment, nor a deterministic result of active, pre-programmed internal factors. So while a firm differs from an industry in being a purposive system, and an industry self-organizes around market exchange, there is some common logic that underpins their evolution. Complexity analysis applied to human activity recognizes that the way a system is perceived and represented shapes motivation and action. In the case of an emerging industry, a sense of identity among participants arises as the system self-organizes but self-organization may in turn promote deliberate forms of organization at the industry level, including industry bodies, trade and employers' institutions. (Garnsey et al., 2014: 179)

This example contracts top-down and bottom-up processes, identifying the fact that bottom-up processes are required when addressing complexity. This adheres to the concept of distributed leadership that encompasses both top-down, bottom-up, and horizontal leadership. Through self-organizing functions, internal structures are capable of altering existing structures to adapt to external perturbations, relating to the concepts of complexity thinking (e.g., emergence, self-organizing, adaptability). The customer first is represented by the idea that the system "self-organizes around market exchange" and team science is represented by the collaborative entities identified (e.g., participants or small groups, organization, industry).

Key Takeaways from The Flow System

In a 2017 "State of the Global Workplace" report, Gallup described that "organizations and institutions have often been slow to adapt to the rapid changes produced by the spread of information technology, the globalization of markets for products and labor, the rise of the fifth economy, and younger workers' unique expectations" (n.a., 2017: 5). This description includes an external source claiming that organizations are failing to adapt to today's complex environment. Following are three general recommendations to repair the global issues pinpointed in the "State of the Global Workplace" report:

- Move the whole world to a workplace strategy of "high development."
- Make every workplace in the world strengths-based.
- Move the world's workplace mission from paycheck to purpose. (Gallup, 2017: 2–3)

TFS easily accomplishes each of these three items. First, development is inherent in the distributed leadership helix with a strong emphasis on leadership development at all levels of the organization. Second, as an organization transitions to a team-based structure, or reorganizes to accommodate existing team-based structures, team composition efforts are based on team member knowledge, skills, abilities, and other characteristics (KSAOs). This optimizes a team's collective ability as required to address the problem or task they have been assigned, maximizing each team member's strengths. Third, one common theme of the distributed leadership concept was that each leader is responsible to instill purpose and meaning in each team and team member's work. This purpose and meaning also should be shared with the customer.

Core Principles and Attributes of The Flow System

To summarize the contents of this book, we next highlight the core principles and attributes of TFS. The core principles of TFS are as follows:

1. Customer First.
2. The *flow* of value.
3. The Triple Helix of Flow.
    a. Complexity Thinking
    b. Distributed Leadership
    c. Team Science

The attributes are divided up among the different core principles of TFS and are provided in the following section.

The Flow System Attributes

TFS-Attribute 1: Provide value to the customer.

TFS-Attribute 2: A nonprescriptive system of understanding the essential components that lead to *flow*.

TFS-Attribute 3: An organization's structure must match its intended function.

TFS-Attribute 4: Organizational strategies must increase the clarity, strength, and presence of weak signals.

TFS-Attribute 5: Weak signals must be acknowledged and addressed.

Customer Value First Attributes

CV-Attribute 1: The purpose of an organization is not to make a profit.

CV-Attribute 2: An organization's purpose is to develop new customers and to maintain existing customers.

Triple Helix of Flow Attributes

TH-Attribute 1: Each of the three components of the triple helix (complexity thinking, distributed leadership, team science) must be interconnected, synchronized, and embedding in an organization's function and structure to achieve flow—delivering value to the customer in complex and disrupted environments.

Complexity Thinking Attributes

CT-Attribute 1: Know when to use systems thinking and when to use complexity thinking tools and techniques.

CT-Attribute 2: Situational Awareness—Know which domain of knowledge (clear, complicated, complex, chaotic, disorder) you are operating in at any given time.

CT-Attribute 3: Know the type of complex environment and/or problem that is being addressed.

CT-Attribute 4: Know the initial steps to complexity thinking.

Step 1 involves understanding the characteristics of complex systems.

Step 2 involves having a worldview or perspective that systems, entities, and events are complex adaptive systems.

CT-Attribute 5: Understand the characteristics of complex systems.

- Complex systems consist of a large number of elements that in themselves can be simple.
- These elements interact dynamically.
- There are many direct and indirect feedback loops.
- Complex systems are open systems.
- Complex systems have memory distributed through the system.
- Complex systems have a history.
- The behavior of the system is determined by the nature of the interactions.
- Complex systems are adaptive. (Cilliers, 2000)

CT-Attribute 6: Understand the characteristics of complex adaptive systems.

- Path Dependency—Systems are sensitive to their initial conditions.
- Systems have a history—Systems have a history that influences how it responds to change.
- Nonlinearity—Systems react to external perturbations disproportionately.
- Emergence—Simple laws of emergence include the following:
  The component mechanisms interact without central control, and
  The possibilities for emergence increase rapidly as the flexibility of the interactions increases. (Holland, 1998: 7)
- Irreducible—Irreversible process transformations cannot be reduced back to their original state.
- Adaptive—Having the capability to process information, to learn, to develop mental models about frozen and unfrozen components of a system, and synthesizing the two, results in being adaptive.

- Operates between order and chaos—Once the behavior of a system is in focus, the extremes of order and chaos can be identified.
- Self-organizing—Systems self-organize through dynamic, interaction, and feedback mechanisms to become adaptive.

CT-Attribute 7: Focuses on what cannot be explained, not what can be explained.

Distributed Leadership Attributes

DL-Attribute 1: Leadership is a shared construct as opposed to an individual construct.

DL-Attribute 2: Leadership is distributed throughout the whole organization; it must extend horizontally, vertically, and everywhere in between.

DL-Attribute 3: Leadership is multidimensional and multilevel.

DL-Attribute 4: Leadership is an emergent outcome that results from interactions.

DL-Attribute 5: Leadership development requires real-world practice.

DL-Attribute 6: Leadership is a shared property of the team or small group in which all team members participate in the leadership process.

DL-Attribute 7: Organizations and institutions with team-based structures need to provide leadership structures designed for team-based structures.

DL-Attribute 8: Shared leadership can be developed.

DL-Attribute 9: Shared leadership and foundational leadership theories include transactional, transformational, laissez-faire, directive, empowering, and functional leadership.

DL-Attribute 10: Hybrid/Integrated Leadership Theories (Strategic, Instrumental, Global leadership) are required for complexity.

DL-Attribute 11: Build the organizational and leadership structure that will be needed for the future.

Team Science Attributes

TS-Attribute 1: Know when teams are needed and when a team is not needed: Teams generally are needed when any one of the following hold true:
a) The task requires more resources than any one person can provide.
b) Diversity of skills and perspectives are required to accomplish the work.
c) Flexibility is needed to keep pace with a rapidly changing context.
d) There is a desire to provide an environment where individual members can hone their personal capabilities through interactions with others. (Hackman, 2011)

TS-Attribute 2: Team processes consist of recurring processes; transition phase processes, action phase processes, and the interpersonal processes.

TS-Attribute 3: Team composition must take into account the taskwork expected of the team, the teamwork and interpersonal relationships expected among team members, and the ability of team members to obtain external resources when needed.

TS-Attribute 4: Taskwork requires team members to have the knowledge, skills, and abilities that enable them to understand the required processes necessary to complete the task and the procedures involved. Some of these activities include orientation, resource distribution, timing, response coordination, motivation, systems monitoring, and procedures maintenance. (Devine, 2002)

TS-Attribute 5: Reciprocal task structures and teams composed divisionally are best for complex environments/problems.

TS-Attribute 6: The requirements of teamwork include, but are limited to, the following:

1. Teamwork means that members monitor one another's performance.
2. Teamwork implies that members provide feedback to and accept it from one another.
3. Teamwork involves effective communication among members, which often involves closed-loop communication.
4. Teamwork implies the willingness, preparedness, and proclivity to back fellow members up during operations.
5. Teamwork involves group members' collectively viewing of themselves as a group whose success depends on their interaction.
6. Teamwork means fostering within-team interdependence.
7. Teamwork is characterized by a flexible repertoire of behavioral skills that vary as a function of circumstances.
8. Teams change over time.
9. Teamwork and taskwork are distinct. (McIntyre and Salas, 1995: 23–33)

TS-Attribute 7: The dimensions of teamwork involve core processes (cooperation, conflict, coordination, communication, coaching, cognition) and influencing behaviors (composition, context, culture). (Dihn and Salas, 2017)

TS-Attribute 8: A team's leadership is a product of the team's culture.

TS-Attribute 9: Team Effectiveness is the point in which team processes are aligned with tasks demands (Kozlowski et al., 2015) and are considered optimized when the processes produce the desired outcome (provide value to the customer). (Driskell et al., 2018)

TS-Attribute 10: The dimensions of teamwork include the following:

1. The productive output of the team (i.e., its product, service, or decision) meets or exceeds the standards of quantity, quality, and timeliness of the team's clients—that is, of the people who receive, review, or use the output.
2. The social processes the team uses in carrying out the work enhance members' capability to work together interdependently in the future.
3. The group experience, on balance, contributes positively to the learning and professional development of individual team members. (Hackman, 2011: 37–39)

TS-Attribute 11: The success of a team can be evaluated by the following:

1. The amount of *effort* members are expending in carrying out their collective work.

2. The task-appropriateness of the team's *performance strategies*, the choices the team makes about how it will carry out the work.

3. The level of *knowledge and skill* the team is applying to the work. (Hackman, 2011: 40)

TS-Attribute 12: TE = (TW + IP) + TK(TP + AP) + PF + CV

Team Effectiveness (TE) is composed of teamwork (TW) as a function of both teamwork and interpersonal phases (IP); taskwork (TK) as a function of transition (TP) and action (AP) phases; performance (PF) composed of quality, quantity and team characteristics; and customer value (CV) as a function of the customers' satisfaction as well as the team member's satisfaction and commitment from their experiences as a team member.

TS-Attribute 13: Team training for teamwork must be done as a team.

TS-Attribute 14: Team effective training must begin in identifying the requisite teamwork skills and competencies that are essential to the current contextual setting.

TS-Attribute 15: Team effectiveness training must include the following:
1. Include the necessary teamwork skills for the contextual setting.
2. Be focused on learning (content and activities) the requisite teamwork skills previously identified.
3. Include team members training together.
4. Must incorporate briefing and debriefing activities.
5. Is best done in situ (in real time).
6. Must be evaluated.

TS-Attribute 16: Multiteam systems (MTS) are structurally formed when two or more teams are working toward a common goal.

TS-Attribute 17: Each component team within an MTS have their own set of team goals (component team proximal goals).

TS-Attribute 18: Component teams within an MTS work toward a shared MTS goal (component team distal goal, MTS superordinate goal).

TS-Attribute 19: Component teams within the same MTS can have multiple and different distal goals from other component teams within the same MTS, but share common distal goals with at least one additional component team within the same MTS.

TS-Attribute 20: Goal hierarchies for MTS entail the following:
- MTS goal hierarchies have a minimum of two levels.
- Goals at higher levels entail greater interdependent actions among more component teams than goals at lower levels.
- The superordinate goal at the apex of the hierarchy rests on the accomplishment by component teams of all lower order goals.
- Higher-order goals are likely to have a longer time horizon than lower-order goals.
- Goals vary in their priority and valence. (Zaccaro et al., 2012)

TS-Attribute 21: Three interdependencies of MTS:
- Input interdependencies: sharing of component teams (human resources, information, technology, resources, capital).
- Process interdependencies: Interactions required between component teams (boundary spanning, environmental sensemaking, task ordering, tactical planning, knowledge transfer, timing, coordination of activities, monitoring and feedback).
- Outcome interdependencies: Component team outcomes as they relate to the MTS superordinate goal.

TS-Attribute 22: MTS effectiveness:

MTS TE (higher level) = C-C + C-BS + Pmts + CVmts + SUM[TE (lower level)]
SUM[TE (lower level)] = Component Team 1 [(TW + IP) + TK(TP + AP) + PF + CV] + ... + Component Team n [(TW + IP) + TK(TP + AP) + PF + CV].

MTS effectiveness requires interactions between component teams as well as interactions between each component team and the boundary spanner. MTS effectiveness requires component team to component team interactions (C-C), component team to boundary spanner interactions (C-BS), performance at the MTS level (Pmts), and customer value at the MTS (CVmts), plus the aggregate of each component team's effectiveness.

TS-Attribute 23: Solving complex problems can be achieved by breaking the problem into multiple levels, spanning "micro and macro scales of the environment that includes individual, small group, organizational, institutional, community, and global levels of analysis" (Stokols, 2018: xxiii–xxiv). MTS provides one means of structuring a multilevel complex problem.

Summary Attribute

TFS-Attribute 6: Implementing MTSs into an organizational structure requires an organizational transformation that includes each of the levels of analysis (individual, team, MTS, organizational), all levels of goal structures (proximal, distal, organizational), and leadership styles (shared leadership at the team level; boundary spanners at the–MTS level; and strategic, instrumental, and global leadership at the organizational level).

References

Ancona D, Backman E, and Isaacs K. (2019) Nimble leadership: Walking the line between creativity and chaos. *Harvard Business Review* 97: 74–83.

Brooks FP. (1995) *The mythical man-month: Essays on software engineering.* Boston, MA: Addison-Wesley.

Cilliers P. (2000) What can we learn from a theory of complexity? *Emergence: Complexity and Organization* 2: 23–33.

Columbia Accident Investigation Board. (2003, August) Report of the Columbia Accident Investigation Board. Washington, DC: NASA.

Devine DJ. (2002) A review and integration of classification systems relevant to teams in organizations. *Group Dynamics: Theory, Research, and Practice* 6: 291–310.

Dihn JV and Salas E. (2017) Factors that influecne teamwork. In: Salas E, Rico R, and Passmore J (eds) *The Wiley Blackwell handbook of the psychology of team working and collaborative processes.* Malden, MA: Wiley, 15–41.

Driskell JE, Salas E, and Driskell T. (2018) Foundations of teamwork and collaboration. *American Psychologist* 73: 334–348.

Drucker PF. (2007) *People and performance: The best of Peter Drucker on management.* Boston, MA: Harvard Business School Press.

Gallup (2017) State of the global workplace. New York, NY: Gallup, Inc.

Garnsey E, Ford S, and Heffernan. (2014) The evolution of industries in diverse markets: A complexity approach. In: Srathern M and McGlade J (eds) *The social face of complexity science.* Litchfield Park, AZ: Emergent Publications, 159–189.

Hackman RJ. (2011) *Collaborative intelligence: Using teams to solve hard problems,* San Francisco, CA: Berrett-Koehler.

Holland JH. (1998) *Emergence: From chaos to order.* Reading, MA: Helix Books.

Isaacson W. (2014) *The innovators: How a group of hackers, geniuses, and geeks created the digital revolution.* New York, NY: Simon & Schuster.

Kozlowski SWJ, Grand JA, Baard SK, et al. (2015) Teams, teamwork, and team effectiveness: Implications for human system integration. In: Boehm-Davis D, Durso F and Lee J (eds) *APA handbook of human systems integration.* Washington, DC: APA, 555–571.

Lewin K. (1945) The Research Center for Group Dynamics at Massachusetts Institute of Technology. *Sociometry* 8: 126–136.

McIntyre RM and Salas E. (1995) Measuring and managing for team performance: Emerging principles from complex environments. In: Guzzo RA and Salas E (eds) *Team effectiveness and decision making in organizations.* San Francisco, CA: Jossey-Bass.

Sarasohn HM and Protzman CA. (1998) *The fundamentals of industrial management: The Homer Satasohn papers, 1936–2001.* Series 1: Civil communications section (CSS), 1946–1998.

Stigler SM. (2010) The changing history of robustness. *American Statistician* 64: 277–281.

Stokols D. (2018) *Social ecology in the digital age: Solving complex problems in a globalized world.* San Diego, CA: Academic Press.

Zaccaro SJ, Marks MA, and Dechurch LA. (2012) Multiteam systems: An introduction. In: Zaccaro SJ, Marks MA, and Dechurch LA (eds) *Multiteam systems: Am organization form for dynamic and complex environments.* New York, NY: Routledge, 3–32.

Zemmel R, Cuddihy M, and Carey D. (2018) How successful CEOs manage their middle act: A strong start takes you only so far. *Harvard Business Review* 96: 98–105.

Index

www.ingramcontent.com/pod-product-compliance
Ingram Content Group UK Ltd.
Pitfield, Milton Keynes, MK11 3LW, UK
UKHW020049070125
453151UK00005B/204

9 798988 023906